STUDENT WORKBOOK

for
Argument-Driven Inquiry
in
Third-Grade Science
Three-Dimensional Investigations

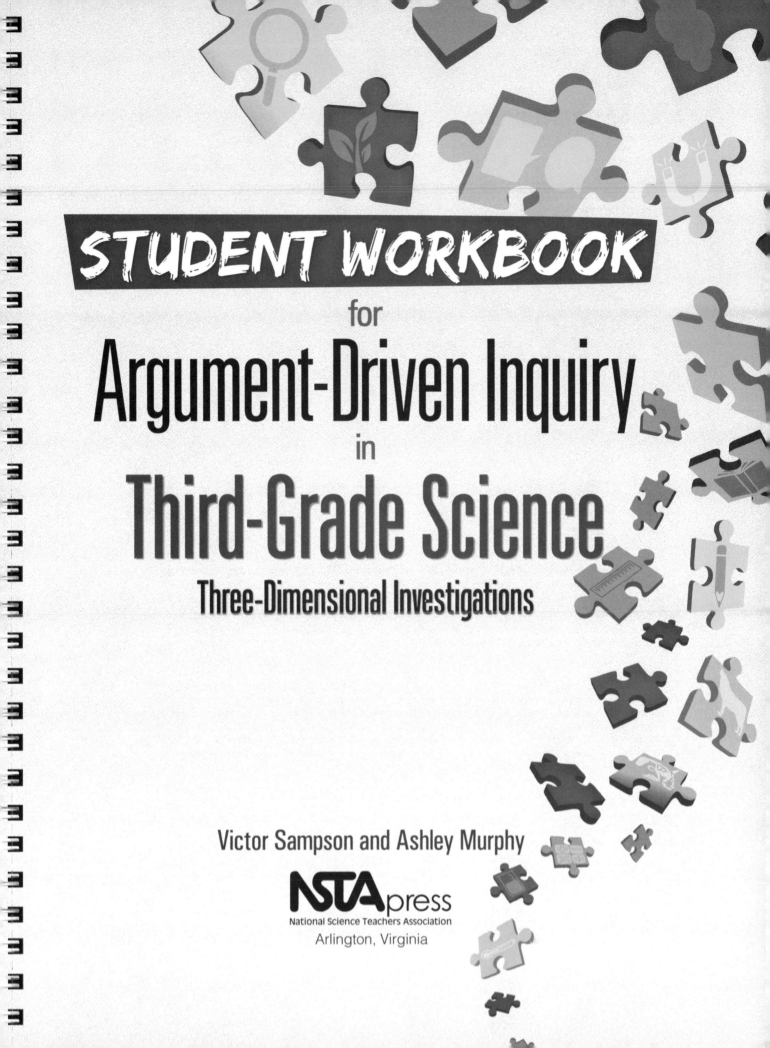

STUDENT WORKBOOK

for
Argument-Driven Inquiry
in
Third-Grade Science

Three-Dimensional Investigations

Victor Sampson and Ashley Murphy

NSTApress
National Science Teachers Association
Arlington, Virginia

National Science Teachers Association

Claire Reinburg, Director
Rachel Ledbetter, Managing Editor
Andrea Silen, Associate Editor
Jennifer Thompson, Associate Editor
Donna Yudkin, Book Acquisitions Manager

ART AND DESIGN
Will Thomas Jr., Director

PRINTING AND PRODUCTION
Catherine Lorrain, Director

NATIONAL SCIENCE TEACHERS ASSOCIATION
David L. Evans, Executive Director

1840 Wilson Blvd., Arlington, VA 22201
www.nsta.org/store
For customer service inquiries, please call 800-277-5300.

Cataloging-in-Publication Data are available from the Library of Congress.
LCCN: 2018041212
ISBN: 978-1-68140-567-4
e-ISBN: 978-1-68140-568-1

Contents

Contents

Contents

Section 1
Introduction and Investigation Safety

Introduction

Science is much more than a collection of facts or terms. Science is a way to figure out how the world works and why it works that way. In science, we use core ideas, crosscutting concepts, and practices to figure things out.

The core ideas of science include the theories and laws that scientists in different fields of science use to develop new explanations for why or how things happen. For example, life scientists use the core idea of heredity to figure out how a trait is passed down from parent to child, and earth scientists use the core idea of weather and climate to figure out why it rains a lot in some places or why it is cooler during different parts of the year.

The crosscutting concepts of science are themes that have value in every field of science as a way to help scientists understand why or how things happen. For example, life scientists and earth scientists both look for patterns and cause-and-effect relationships when they are trying to develop explanations for why or how things happen. These crosscutting concepts help them think about what it is important to think about or look for during an investigation.

Finally, scientists use the practices of science to develop and refine new ideas. The practices include such things as asking and answering questions, planning and carrying out investigations, analyzing and interpreting data, and obtaining, evaluating, and communicating information. One of the most important practices of science is arguing from evidence. These core ideas, crosscutting concepts, and practices of science are important in science because most, if not all, scientists use them to figure out how the world works and why it works that way.

These core ideas, crosscutting concepts, and scientific practices are important for you to learn while you are in school. When you understand these core ideas, crosscutting concepts, and practices, it is easier to make sense of what scientists try to do. It is also easier to talk about what is currently going on in science with other people and to evaluate what you read or hear about science in the news. Once you learn the core ideas, crosscutting concepts, and practices of science, you will also have the knowledge and skills that you need to continue to learn about science outside school or to enter a career in science, technology, or engineering.

The investigations that are included in this book are designed to help you learn the core ideas, crosscutting concepts, and practices of science. During each investigation, you will have an opportunity to use a core idea, a crosscutting concept, and several practices of science to figure something out. Your teacher will introduce each investigation by giving you a task to accomplish and a guiding question to answer. You will then work as part of a group to plan and carry out an investigation to collect the data that you need to answer that question. From there, your group will develop an argument that includes an answer to the guiding question. You will then have an opportunity to share your argument with your classmates and critique their arguments, much like professional scientists do.

Next, you will be asked to revise your draft argument based on their feedback. You will then be asked to write an investigation report on your own to share what you learned. Your classmates will review this report before you submit it to your teacher for a grade.

As you complete more and more investigations in this book, you will not only learn the core ideas associated with each investigation but will also get better at using the crosscutting concepts and practices of science to understand the natural world.

Safety Rules

You will be doing many different investigations this year. You will use different materials, tools, and chemicals during these investigations. It is very important that you follow these 11 safety rules to keep you and your classmates from getting hurt when you use them:

1. Act in a responsible manner at all times.
2. Do not eat, drink, or chew gum.
3. Do not touch, taste, or smell any materials, tools, or chemicals without permission.
4. Wear safety goggles at all times.
5. Take care of the materials and tools that you use.
6. Tell your teacher about any accidents as soon as they happen.
7. Wear appropriate clothing, including closed-toed shoes and pants. Clothes should not be loose, baggy, or bulky. Use hair ties to keep long hair out of the way.
8. Keep work areas clean and neat at all times.
9. Clean work area and any materials or tools at the end of the investigations.
10. Wash your hands with soap and water at the end of the investigations.
11. Follow the teacher's directions at all times.

Your teacher may provide some more rules to follow during some of your investigations, but you must always follow these 11 rules. Your teacher will also warn you about anything that you need to keep in mind to stay safe as you work though an investigation.

Your teacher will go over a safety acknowledgment form with you before you start your first investigation. You will need to sign this safety acknowledgment form so your teacher knows that you understand all the safety rules and agree to follow them. Be sure to show this form to a parent or guardian after you have signed it. Your parent or guardian will also need to read and sign the safety acknowledgment form before your teacher will allow you to participate in any investigations.

Safety Acknowledgment Form

I know that it is very important to be as safe as I can during an investigation. My teacher has told me how to be safer in science. I agree to follow these 11 safety rules when I am working with my classmates to figure things out in science:

1. I will act in a responsible manner at all times. I will not run around the classroom, throw things, play jokes on my classmates, or be careless.
2. I will never eat, drink, or chew gum.
3. I will never touch, taste, or smell any materials, tools, or chemicals without permission.
4. I will wear my safety goggles at all times during the activity setup, hands-on work, and cleanup.
5. I will do my best to take care of the materials and tools that my teacher allows me to use.
6. I will always tell my teacher about any accidents as soon as they happen.
7. I will always dress in a way that will help keep me safer. I will wear closed-toed shoes and pants. My clothes will not be loose, baggy, or bulky. I will also use hair ties to keep my hair out of the way while I am working if my hair is long.
8. I will keep my work area clean and neat at all times. I will put my backpack, books, and other personal items where my teacher tells me to put them and I will not get them out unless my teacher tells me that it is okay.
9. I will clean my work area and the materials or tools that I use.
10. I will wash my hands with soap and water at the end of the activity.
11. I will follow my teacher's directions at all times.

_____   _____   _____
           Print Name   Signature   Date

I have read and reviewed the 11 investigation safety rules with my child. He or she understands how important it is to follow safety rules in science and has agreed to follow these safety rules at all times. I give my permission for my child to participate in the investigations this year.

_____   _____   _____
   Parent or Guardian Name   Parent or Guardian Signature   Date

Section 2
Motion and Stability: Forces and Interactions

Investigation 1

Magnetic Attraction: What Types of Objects Are Attracted to a Magnet?

Introduction

We use magnets every day. We hold pictures up on the refrigerator with magnets. We also use magnets to keep the refrigerator door shut. There are even magnets in many different kinds of toys. Take a few minutes to explore what happens when you bring a magnet near a bunch of paper clips. Keep track of what you observe and what you are wondering about in the boxes below.

Things I OBSERVED …	Things I WONDER about …

Magnets come in different shapes and sizes. Most magnets are made of iron. Magnets are amazing because they can stick to an object. Magnets can also make an object move without touching it. Scientists call the force produced by a magnet that can make things stick to it or that makes an object move without touching it *magnetism*.

You have probably seen magnets stick to some objects but not others. Magnets do not stick to all objects because objects are made of different materials. For example, beverage cans are made out of aluminum. We use wood and graphite to make pencils. Wires are made out of copper and plastic. These different materials have different *physical properties*. An example of a physical property is color. The ability to stick to a magnet is another example of a physical property. Materials are different from each other because they have different physical properties.

Your goal in this investigation is to *figure out* what types of objects are attracted to a magnet and what types of objects are not attracted to a magnet. To complete this task, you will need to test different objects to see if they stick to a magnet or not. You will then need to look for patterns in the objects that are attracted to magnets and the objects that are not attracted to magnets. Scientists often look for patterns in nature like this. They then use these patterns to classify objects or sort objects into groups. You can use the patterns you find to figure out what is similar about objects that are attracted to a magnet and what makes them different from objects that are not attracted to a magnet.

Things we KNOW from what we read …	What we will NEED to figure out …

Your Task

Use what you know about magnetism, properties of matter, and patterns to design and carry out an investigation to determine what types of materials are attracted to a magnet and what types of materials are not.

The *guiding question* of this investigation is, **What types of objects are attracted to a magnet?**

Materials

You may use any of the following materials during your investigation:

- Safety glasses or goggles (required)
- Ceramic magnet (iron)
- Foil (aluminum)
- Wire (copper)
- Penny (zinc and copper)

- Nickel (copper and nickel)
- Nail (iron)
- Paper clips (iron)
- BBs (iron)
- BBs (plastic)
- Washer (zinc and iron)

- Fishing sinker (lead)
- String (cotton)
- Index card (paper)
- Tile (plastic)
- Tile (glass)
- Marble (glass)
- Block (wood)

Safety Rules

Follow all normal safety rules. In addition, be sure to follow these rules:

- Wear sanitized safety glasses or goggles during setup, investigation activity, and cleanup.
- Do not throw objects or put any objects in your mouth.
- Use caution when handling sharp objects (e.g., wire, nails, paper clips). They can cut or puncture skin.
- Immediately pick up any slip or fall hazards (e.g., marbles) from the floor.
- Wash your hands with soap and water when you are done collecting the data.

Plan Your Investigation

Prepare a plan for your investigation by filling out the chart that follows; this plan is called an *investigation proposal*. Before you start developing your plan, be sure to discuss the following questions with the other members of your group:

- What information should we collect so we can **compare and contrast** different materials?
- What types of **patterns** might we look for to help answer the guiding question?

Investigation Log

Our guiding question:

We will collect the following data:

These are the steps we will follow to collect data:

I approve of this investigation proposal.

_____   _____
Teacher's signature   Date

National Science Teachers Association

Collect Your Data

Keep a record of what you measure or observe during your investigation in the space below.

Analyze Your Data

You will need to analyze the data you collected before you can develop an answer to the guiding question. In the space below, create a table or figure that shows how objects that are attracted to a magnet are similar to each other and different from objects that are not attracted to a magnet.

Draft Argument

Develop an argument on a whiteboard. It should include the following parts:

1. A *claim:* Your answer to the guiding question.

2. *Evidence:* An analysis of the data and an explanation of what the analysis means.

3. A *justification of the evidence:* Why your group thinks the evidence is important.

The Guiding Question:	
Our Claim:	
Our Evidence:	Our Justification of the Evidence:

Argumentation Session

Share your argument with your classmates. Be sure to ask them how to make your draft argument better. Keep track of their suggestions in the space below.

Ways to IMPROVE our argument …

Draft Report

Prepare an *investigation report* to share what you have learned. Use the information in this handout and your group's final argument to write a *draft* of your investigation report.

Introduction

We have been studying _____ in class.

Before we started this investigation, we explored _____

_____

_____

_____

_____

We noticed _____

_____

_____

_____

_____

My goal for this investigation was to figure out _____

_____

_____

_____

_____

The guiding question was _____

_____

_____

_____

_____

Investigation Log

Method

To gather the data I needed to answer this question, I _____

_____

_____

_____

_____

_____

_____

I then analyzed the data I collected by _____

_____

_____

_____

_____

_____

_____

_____

_____

Argument

My claim is _____

_____

_____

_____

_____

_____

National Science Teachers Association

The _____ below shows _____

_____

_____

This evidence is important because _____

_____

_____

_____

_____

 **Review**

Your friends need your help! Review the draft of their investigation reports and
give them ideas about how to improve. Use the *peer-review guide* that begins on the
next page to guide your review.

Investigation Log

Peer-Review Guide

Section 1: The Investigation	Reviewer Rating		
1. Did the author do a good job of explaining what the investigation was about?	☐ No	☐ Almost	☐ Yes
2. Did the author do a good job of making the **guiding question** clear?	☐ No	☐ Almost	☐ Yes
3. Did the author do a good job of describing what he or she did to **collect data?**	☐ No	☐ Almost	☐ Yes
4. Did the author do a good job describing **how** he or she **analyzed** the data?	☐ No	☐ Almost	☐ Yes

Reviewers: If your group gave the author any "No" or "Almost" ratings, please give the author some advice about what to do to improve this part of his or her investigation report.

Section 2: The Argument	Reviewer Rating		
1. Does the author's claim provide a clear and detailed **answer** to the guiding question?	☐ No	☐ Almost	☐ Yes
2. Did the author support his or her claim with **scientific evidence?** Scientific evidence includes analyzed data and an explanation of the analysis.	☐ No	☐ Almost	☐ Yes
3. Does the **evidence** that the author uses in his or her argument **support the claim?**	☐ No	☐ Almost	☐ Yes
4. Did the author include enough **evidence** in his or her argument?	☐ No	☐ Almost	☐ Yes
5. Did the author do a good job of **explaining why the evidence** is important (why it matters)?	☐ No	☐ Almost	☐ Yes
6. Is the content of the argument **correct** based on the science concepts we talked about in class?	☐ No	☐ Almost	☐ Yes

Reviewers: If your group gave the author any "No" or "Almost" ratings, please give the author some advice about what to do to improve this part of his or her investigation report.

Continued

National Science Teachers Association

Section 3: Mechanics	Reviewer Rating		
1. ***Grammar:*** Are the sentences complete? Is there proper subject-verb agreement in each sentence? Are there no run-on sentences?	☐ No	☐ Almost	☐ Yes
2. ***Conventions:*** Did the author use proper spelling, punctuation, and capitalization?	☐ No	☐ Almost	☐ Yes
3. ***Word Choice:*** Did the author use the right words in each sentence (for example, *there* vs. *their, to* vs. *too, then* vs. *than*)?	☐ No	☐ Almost	☐ Yes

Reviewers: If your group gave the author any "No" or "Almost" ratings, please give the author some advice about what to do to improve the writing mechanics of his or her investigation report.

General Reviewer Comments

We liked …

We wonder …

Write Your Final Report

Once you have received feedback from your friends about your draft report, create your final investigation report in the space that follows.

Introduction

_____

_____

_____

_____

_____

_____

_____

Method

_____

_____

_____

_____

_____

_____

_____

_____

_____

_____

_____

Argument

_____

_____

_____

_____

_____

_____

_____

_____

_____

_____

_____

_____

Investigation Log

Investigation Report Grading Rubric

Section 1: The Investigation	Missing	Somewhat	Yes
		Score	
1. The author explained what the investigation was about.	0	1	2
2. The author made the **guiding question** clear.	0	1	2
3. The author **described** what he or she did to **collect data.**	0	1	2
4. The author described **how** he or she **analyzed** the data.	0	1	2

Section 2: The Argument	Missing	Somewhat	Yes
		Score	
1. The claim includes a clear and detailed **answer** to the guiding question.	0	1	2
2. The author used **scientific evidence** to support the claim. Scientific evidence includes analyzed data and an explanation of the analysis.	0	1	2
3. The evidence **supports the claim.**	0	1	2
4. The author included enough **evidence** in his or her argument.	0	1	2
5. The author **explained why the evidence** is important.	0	1	2
6. The content of the argument is **correct.**	0	1	2

Section 3: Mechanics	Missing	Somewhat	Yes
		Score	
1. *Grammar:* The sentences are complete. There is proper subject-verb agreement in each sentence. There are no run-on sentences.	0	1	2
2. *Conventions:* The author used proper spelling, punctuation, and capitalization.	0	1	2
3. *Word Choice:* The author used the right words in each sentence (e.g., *there* vs. *their, to* vs. *too, then* vs. *than*).	0	1	2

Teacher Comments

Here are some things I really liked about your report …	Here are some things I think you could do next time to make your report even better …

Total: _____ /26

National Science Teachers Association

Checkout Questions

Investigation 1.
Magnetic Attraction: What Types of Objects Are Attracted to a Magnet?

1. Listed below are some objects. Place an *X* next to the objects that will stick to a magnet.

☐ Book	☐ Paper clip	☐ Piece of tin
☐ Coffee cup	☐ Piece of aluminum foil	☐ Plastic toy
☐ Nail	☐ Piece of copper wire	☐ Seed
☐ Nickel	☐ Piece of paper	☐ Wood block

2. Explain your thinking. What *pattern* from your investigation did you use to decide if an object will or will not stick to a magnet?

_____

_____

_____

_____

_____

_____

_____

_____

Teacher Scoring Rubric for the Checkout Questions

Level	Description
3	The student can apply the core idea correctly in all cases and can fully explain the pattern.
2	The student can apply the core idea correctly in all cases but cannot fully explain the pattern.
1	The student cannot apply the core idea correctly in all cases but can fully explain the pattern.
0	The student cannot apply the core idea correctly in all cases and cannot explain the pattern.

Investigation 2

Magnetic Force: How Does Changing the Distance Between Two Magnets Affect Magnetic Force Strength?

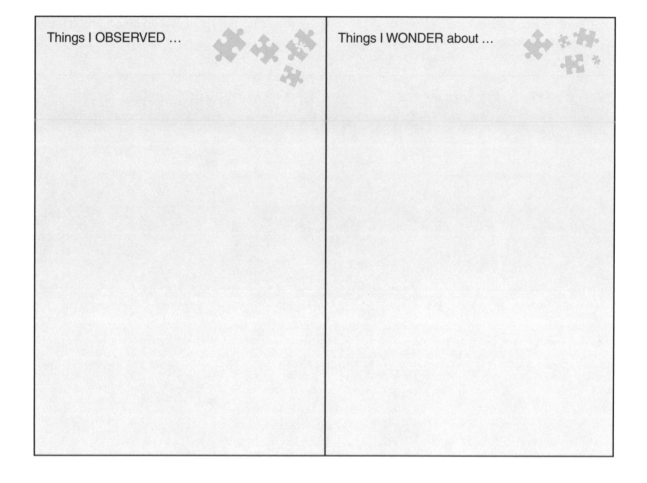

Introduction

Magnets are very useful. We can use a magnet to pull objects that are made of iron or contain iron closer to us without touching them. We can also use magnets to hold things in place because the force of a magnet can travel through objects. For example, we can use a magnet to hold a piece of paper on a refrigerator. Magnets can also push or pull on other magnets. Take a moment to explore what happens when you bring one magnet near another magnet. Keep track of what you observe and what you are wondering about in the boxes below.

Things I OBSERVED ...	Things I WONDER about ...

There is a *magnetic field* around every magnet. We cannot see the magnetic field, but it can apply a push or pull on another magnet. Scientists call this push or pull a *non-contact* force because it does not require the two magnets to be touching each other. The magnetic field is strongest around the two ends of a magnet. The ends of a magnet are called poles. One end of the magnet is called the *north pole,* and the other end of the magnet is called the *south pole.*

The two poles of a magnet may look the same, but they do not act the same. If you put the south pole of a magnet near the north pole of a second magnet, the two magnets will pull together (attract). If you put the south pole of a magnet near the south pole of a second magnet, or if you put the north pole of one magnet near the north pole of a different magnet, the two magnets will push away from each other (repel). In all magnets, identical poles repel each other and different poles attract.

The magnetic field surrounding a magnet is not very large. That is why you must bring two magnets close together for the two magnets to attract or repel. Strong magnets have larger magnetic fields than weak magnets. The magnetic field around a strong magnet will also cause a greater push or pull than the push or pull that is caused by a weak magnet.

Your goal in this investigation is to figure out how changing the distance between two magnets (a cause) changes or does not change the strength of the magnetic force between the two magnets (the effect). You can figure out the strength of the magnetic force between the two magnets by measuring how far one magnet moves away from the other magnet when the two magnets are placed near each other. You can figure out how strong the magnetic force is between two magnets by measuring how far one magnet moves away from another magnet for two reasons. The first reason is that a non-contact force such as magnetism can cause an object to move. The second reason is that an object will travel a greater distance when more force is applied to it.

Things we KNOW from what we read …	What we will NEED to figure out …

Your Task

Use what you know about magnets and cause and effect to design and carry out an investigation to determine how the strength of the magnetic force between two magnets changes as the distance between them changes.

The *guiding question* of this investigation is, **How does changing the distance between two magnets affect magnetic force strength?**

Materials

You may use any of the following materials during your investigation:

- Safety glasses or goggles (required)
- 2 Alnico cylinder magnets
- Ruler with groove

Safety Rules

Follow all normal safety rules. In addition, be sure to follow these rules:

- Wear sanitized safety glasses or goggles during setup, investigation activity, and cleanup.
- Do not throw objects or put any objects in your mouth.
- Wash your hands with soap and water when you are done collecting the data.

Plan Your Investigation

Prepare a plan for your investigation by filling out the chart that follows; this plan is called an *investigation proposal*. Before you start developing your plan, be sure to discuss the following questions with the other members of your group:

- What types of **patterns** might we look for to help answer the guiding question?
- What information do we need to find a **cause-and-effect relationship**?

Our guiding question:

This is a picture of how we will set up the equipment:

We will collect the following data:

These are the steps we will follow to collect data:

I approve of this investigation proposal.

_____   _____
Teacher's signature   Date

Collect Your Data

Keep a record of what you measure or observe during your investigation in the space below.

Analyze Your Data

You will need to analyze the data you collected before you can develop an answer to the guiding question. To do this, create a graph that shows the relationship between the cause and the effect.

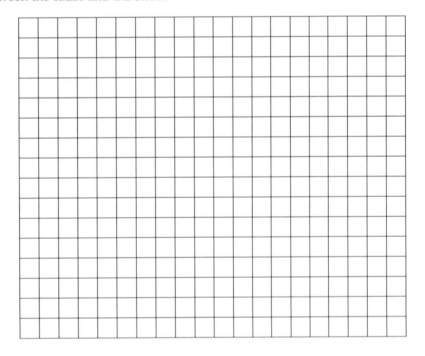

Investigation Log

Draft Argument

Develop an argument on a whiteboard. It should include the following parts:

1. A *claim:* Your answer to the guiding question.

2. *Evidence:* An analysis of the data and an explanation of what the analysis means.

3. A *justification of the evidence:* Why your group thinks the evidence is important.

The Guiding Question:	
Our Claim:	
Our Evidence:	Our Justification of the Evidence:

Argumentation Session

Share your argument with your classmates. Be sure to ask them how to make your draft argument better. Keep track of their suggestions in the space below.

Ways to IMPROVE our argument …

National Science Teachers Association

Draft Report

Prepare an *investigation report* to share what you have learned. Use the information in this handout and your group's final argument to write a *draft* of your investigation report.

Introduction

We have been studying _____ in class.

Before we started this investigation, we explored _____

_____

_____

We noticed _____

_____

_____

My goal for this investigation was to figure out _____

_____

_____

The guiding question was _____

_____

_____

Method

To gather the data I needed to answer this question, I _____

_____

_____

_____

_____

_____

Investigation Log

I then analyzed the data I collected by _____

_____

_____

_____

Argument

My claim is _____

_____

_____

_____

The graph below shows _____

_____

_____

_____

_____

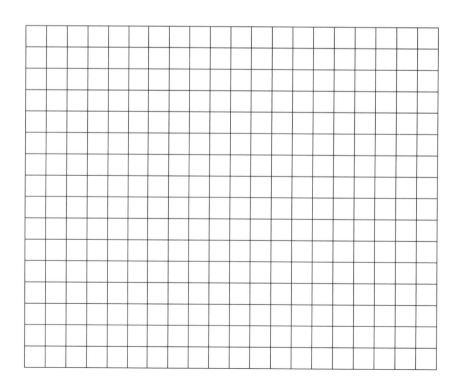

National Science Teachers Association

This evidence is important because _____

_____

_____

_____

_____

_____

_____

_____

 **Review**

Your friends need your help! Review the draft of their investigation reports and give them ideas about how to improve. Use the *peer-review guide* that begins on the next page to guide your review.

Investigation Log

Peer-Review Guide

Section 1: The Investigation	Reviewer Rating		
1. Did the author do a good job of explaining what the investigation was about?	☐ No	☐ Almost	☐ Yes
2. Did the author do a good job of making the **guiding question** clear?	☐ No	☐ Almost	☐ Yes
3. Did the author do a good job of describing what he or she did to **collect data?**	☐ No	☐ Almost	☐ Yes
4. Did the author do a good job describing **how** he or she **analyzed** the data?	☐ No	☐ Almost	☐ Yes
Reviewers: If your group gave the author any "No" or "Almost" ratings, please give the author some advice about what to do to improve this part of his or her investigation report.			

Section 2: The Argument	Reviewer Rating		
1. Does the author's claim provide a clear and detailed **answer** to the guiding question?	☐ No	☐ Almost	☐ Yes
2. Did the author support his or her claim with **scientific evidence?** Scientific evidence includes analyzed data and an explanation of the analysis.	☐ No	☐ Almost	☐ Yes
3. Does the **evidence** that the author uses in his or her argument **support the claim?**	☐ No	☐ Almost	☐ Yes
4. Did the author include enough **evidence** in his or her argument?	☐ No	☐ Almost	☐ Yes
5. Did the author do a good job of **explaining why the evidence** is important (why it matters)?	☐ No	☐ Almost	☐ Yes
6. Is the content of the argument **correct** based on the science concepts we talked about in class?	☐ No	☐ Almost	☐ Yes
Reviewers: If your group gave the author any "No" or "Almost" ratings, please give the author some advice about what to do to improve this part of his or her investigation report.			

Continued

Section 3: Mechanics	Reviewer Rating		
1. *Grammar:* Are the sentences complete? Is there proper subject-verb agreement in each sentence? Are there no run-on sentences?	☐ No	☐ Almost	☐ Yes
2. *Conventions:* Did the author use proper spelling, punctuation, and capitalization?	☐ No	☐ Almost	☐ Yes
3. *Word Choice:* Did the author use the right words in each sentence (for example, *there* vs. *their, to* vs. *too, then* vs. *than*)?	☐ No	☐ Almost	☐ Yes

Reviewers: If your group gave the author any "No" or "Almost" ratings, please give the author some advice about what to do to improve the writing mechanics of his or her investigation report.

General Reviewer Comments

We liked …

We wonder …

Investigation Log

Write Your Final Report

Once you have received feedback from your friends about your draft report, create your final investigation report in the space that follows.

Introduction

_____

_____

_____

_____

_____

_____

_____

Method

_____

_____

_____

_____

_____

_____

_____

_____

_____

_____

_____

National Science Teachers Association

Argument

_____

_____

_____

_____

_____

_____

_____

_____

_____

_____

_____

Investigation Report Grading Rubric

Section 1: The Investigation	Score		
	Missing	Somewhat	Yes
1. The author explained what the investigation was about.	0	1	2
2. The author made the **guiding question** clear.	0	1	2
3. The author **described** what he or she did to **collect data.**	0	1	2
4. The author described **how** he or she **analyzed** the data.	0	1	2

Section 2: The Argument	Score		
	Missing	Somewhat	Yes
1. The claim includes a clear and detailed **answer** to the guiding question.	0	1	2
2. The author used **scientific evidence** to support the claim. Scientific evidence includes analyzed data and an explanation of the analysis.	0	1	2
3. The evidence **supports the claim.**	0	1	2
4. The author included enough **evidence** in his or her argument.	0	1	2
5. The author **explained why the evidence** is important.	0	1	2
6. The content of the argument is **correct.**	0	1	2

Section 3: Mechanics	Score		
	Missing	Somewhat	Yes
1. *Grammar:* The sentences are complete. There is proper subject-verb agreement in each sentence. There are no run-on sentences.	0	1	2
2. *Conventions:* The author used proper spelling, punctuation, and capitalization.	0	1	2
3. *Word Choice:* The author used the right words in each sentence (e.g., *there* vs. *their, to* vs. *too, then* vs. *than*).	0	1	2

Teacher Comments

Here are some things I really liked about your report …	Here are some things I think you could do next time to make your report even better …

Total: _____ /26

National Science Teachers Association

Checkout Questions

Investigation 2. Magnetic Force: How Does Changing the Distance Between Two Magnets Affect Magnetic Force Strength?

1. Listed below are four sets of magnets. Rank the sets of magnets by the strength of the *repelling* force that exists between them. Use a 1 for the *strongest* force and use a 4 for the *weakest* force.

Magnets	**Rank**

A. [N   S]   [S   N]   _____

B. [N   S] [S   N]   _____

C. [S   N]   [N   S]   _____

D. [S   N] [N   S]   _____

2. Explain your thinking. What *cause-and-effect relationship* did you use to rank the strength of the repelling force between the magnets?

_____

_____

_____

_____

_____

_____

Teacher Scoring Rubric for the Checkout Questions

Level	Description
3	The student can apply the core idea correctly in all cases and can fully explain the cause-and-effect relationship.
2	The student can apply the core idea correctly in all cases but cannot fully explain the cause-and-effect relationship.
1	The student cannot apply the core idea correctly in all cases but can fully explain the cause-and-effect relationship.
0	The student cannot apply the core idea correctly in all cases and cannot explain the cause-and-effect relationship.

Investigation 3

Changes in Motion: Where Will the Marble Be Located Each Time It Changes Direction in a Half-Pipe?

Introduction

Objects often move in a predictable pattern. For example, when we bounce a ball, it moves up and down in a predictable pattern. Children playing on swings or seesaws also follow a predictable pattern as they move back and forth or up and down. Take a moment to explore what happens when you place a marble in a half-pipe. Be sure to place the marble at different spots in the half-pipe and pay attention to how the marble moves after each time you let go of it. As you use these materials, keep track of what you observe and what you are wondering about in the boxes below.

Things I OBSERVED ...	Things I WONDER about ...

Investigation Log

The motion of the marble in a half-pipe is another example of an object moving in a predictable pattern. Once a scientist discovers a pattern in the way an object moves, he or she can describe the motion of that object using numbers. This is important because scientists can predict the future motion of an object when they can use numbers to describe the pattern an object follows as it moves.

To be able to describe the motion of an object using numbers, a scientist needs to be able to measure how far an object moves from its initial position. Scientists call this measurement the *displacement* of an object. To measure the displacement of an object, a scientist must pick a reference point. He or she can then measure how far an object moves from this reference point in a given direction. A scientist can also record how long it takes for an object to move from one position to a different position. Scientists call this measurement the *speed* of an object. Speed is how fast an object travels. With this information, a scientist can use numbers to describe the motion of an object by reporting the direction an object moved (left, right, up, or down), how far it moved (its displacement), how long it took for the object to reach the new position (time), and its speed (how fast it moved).

In this investigation, you will examine how a marble moves back and forth in a half-pipe. Your goal is to find a way to describe the marble's motion using numbers. The half-pipe track has ruler tape attached to it to make it easier for you to measure how far the marble moves from the reference point on the track. The reference point is located at the middle of the half-pipe. It is marked with a circle. You will also have a stopwatch so you can keep track of time. With this information, you should be able to not only describe how the marble moves using numbers but also use this pattern to predict the motion of the marble when it is placed in different starting positions in the half-pipe.

Things we KNOW from what we read …	What we will NEED to figure out …

Your Task

Use what you know about the motion of objects and patterns to design and carry out an investigation to describe the pattern a marble follows as it rolls back and forth in a half-pipe. To create the pattern, you will need to use numbers to describe how the marble changes position over time. You will then need to show that you can use the pattern you developed to make accurate predictions about how the marble will move when it is placed in different starting positions in the half-pipe.

The *guiding question* of this investigation is, **Where will the marble be located each time it changes direction in a half-pipe?**

Materials

You may use any of the following materials during your investigation:

- Safety glasses or goggles (required)
- 3 plastic tracks
- 1 marble
- Stopwatch
- Ruler tape

Safety Rules

Follow all normal safety rules. In addition, be sure to follow these rules:

- Wear sanitized safety glasses or goggles during setup, investigation activity, and cleanup.
- Do not throw objects or put any objects your mouth.
- Immediately pick up any slip or fall hazards (e.g., marbles) from the floor.
- Wash your hands with soap and water when you are done collecting the data.

Plan Your Investigation

Prepare a plan for your investigation by filling out the chart that follows; this plan is called an *investigation proposal*. Before you start developing your plan, be sure to discuss the following questions with the other members of your group:

- What information should we collect so we can **describe** the motion of the ball?
- What types of **patterns** might we look for to help answer the guiding question?

Our guiding question:

This is a picture of how we will set up the equipment:

We will collect the following data:

These are the steps we will follow to collect data:

I approve of this investigation proposal.

_____   _____
Teacher's signature   Date

National Science Teachers Association

Collect Your Data

Keep a record of what you measure or observe during your investigation in the space below.

Analyze Your Data

You will need to analyze the data you collected before you can develop an answer to the guiding question. To do this, create a graph that shows how the position of the marble (distance from the reference point) changed as it went back and forth in the half-pipe.

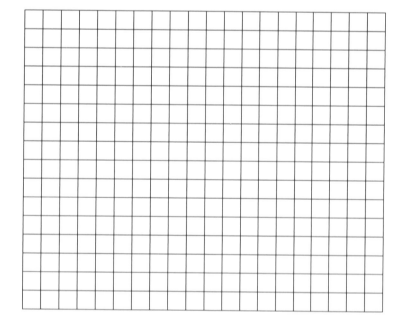

Draft Argument

Develop an argument on a whiteboard. It should include the following parts:

1. A *claim:* Your answer to the guiding question.

2. *Evidence:* An analysis of the data and an explanation of what the analysis means.

3. A *justification of the evidence:* Why your group thinks the evidence is important.

The Guiding Question:	
Our Claim:	
Our Evidence:	Our Justification of the Evidence:

Argumentation Session

Share your argument with your classmates. Be sure to ask them how to make your draft argument better. Keep track of their suggestions in the space below.

Ways to IMPROVE our argument …

Draft Report

Prepare an *investigation report* to share what you have learned. Use the information in this handout and your group's final argument to write a *draft* of your investigation report.

Introduction

We have been studying _____ in class.

Before we started this investigation, we explored _____

_____

We noticed _____

_____

_____

My goal for this investigation was to figure out _____

_____

_____

The guiding question was _____

_____

_____

Method

To gather the data I needed to answer this question, I _____

_____

_____

_____

_____

_____

I then analyzed the data I collected by _____

_____

_____

Argument

My claim is _____

_____

_____

The figure below shows _____

_____

_____

_____

_____

_____

_____

_____

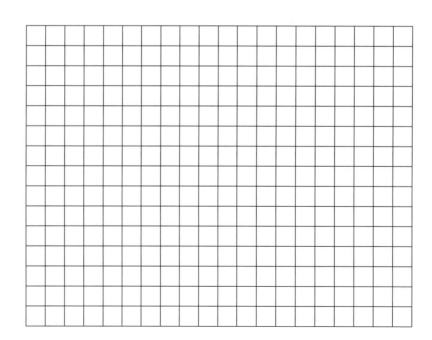

National Science Teachers Association

This evidence is important because _____

_____

_____

_____

_____

_____

_____

_____

 **Review**

Your friends need your help! Review the draft of their investigation reports and give them ideas about how to improve. Use the *peer-review guide* that begins on the next page to guide your review.

Investigation Log

Peer-Review Guide

Section 1: The Investigation	Reviewer Rating		
1. Did the author do a good job of explaining what the investigation was about?	☐ No	☐ Almost	☐ Yes
2. Did the author do a good job of making the **guiding question** clear?	☐ No	☐ Almost	☐ Yes
3. Did the author do a good job of describing what he or she did to **collect data?**	☐ No	☐ Almost	☐ Yes
4. Did the author do a good job describing **how** he or she **analyzed** the data?	☐ No	☐ Almost	☐ Yes

Reviewers: If your group gave the author any "No" or "Almost" ratings, please give the author some advice about what to do to improve this part of his or her investigation report.

Section 2: The Argument	Reviewer Rating		
1. Does the author's claim provide a clear and detailed **answer** to the guiding question?	☐ No	☐ Almost	☐ Yes
2. Did the author support his or her claim with **scientific evidence?** Scientific evidence includes analyzed data and an explanation of the analysis.	☐ No	☐ Almost	☐ Yes
3. Does the **evidence** that the author uses in his or her argument **support the claim?**	☐ No	☐ Almost	☐ Yes
4. Did the author include enough **evidence** in his or her argument?	☐ No	☐ Almost	☐ Yes
5. Did the author do a good job of **explaining why the evidence** is important (why it matters)?	☐ No	☐ Almost	☐ Yes
6. Is the content of the argument **correct** based on the science concepts we talked about in class?	☐ No	☐ Almost	☐ Yes

Reviewers: If your group gave the author any "No" or "Almost" ratings, please give the author some advice about what to do to improve this part of his or her investigation report.

Continued

National Science Teachers Association

Section 3: Mechanics	Reviewer Rating		
1. *Grammar:* Are the sentences complete? Is there proper subject-verb agreement in each sentence? Are there no run-on sentences?	☐ No	☐ Almost	☐ Yes
2. *Conventions:* Did the author use proper spelling, punctuation, and capitalization?	☐ No	☐ Almost	☐ Yes
3. *Word Choice:* Did the author use the right words in each sentence (for example, *there* vs. *their, to* vs. *too, then* vs. *than*)?	☐ No	☐ Almost	☐ Yes

Reviewers: If your group gave the author any "No" or "Almost" ratings, please give the author some advice about what to do to improve the writing mechanics of his or her investigation report.

General Reviewer Comments

We liked …

We wonder …

Write Your Final Report

Once you have received feedback from your friends about your draft report, create your final investigation report in the space that follows.

Introduction

_____

_____

_____

_____

_____

_____

_____

Method

_____

_____

_____

_____

_____

_____

_____

_____

_____

_____

_____

Argument

_____

_____

_____

_____

_____

_____

_____

_____

_____

_____

_____

_____

Investigation Report Grading Rubric

Section 1: The Investigation	Score		
	Missing	Somewhat	Yes
1. The author explained what the investigation was about.	0	1	2
2. The author made the **guiding question** clear.	0	1	2
3. The author **described** what he or she did to **collect data.**	0	1	2
4. The author described **how** he or she **analyzed** the data.	0	1	2

Section 2: The Argument	Score		
	Missing	Somewhat	Yes
1. The claim includes a clear and detailed **answer** to the guiding question.	0	1	2
2. The author used **scientific evidence** to support the claim. Scientific evidence includes analyzed data and an explanation of the analysis.	0	1	2
3. The evidence **supports the claim.**	0	1	2
4. The author included enough **evidence** in his or her argument.	0	1	2
5. The author **explained why the evidence** is important.	0	1	2
6. The content of the argument is **correct.**	0	1	2

Section 3: Mechanics	Score		
	Missing	Somewhat	Yes
1. *Grammar:* The sentences are complete. There is proper subject-verb agreement in each sentence. There are no run-on sentences.	0	1	2
2. *Conventions:* The author used proper spelling, punctuation, and capitalization.	0	1	2
3. *Word Choice:* The author used the right words in each sentence (e.g., *there* vs. *their, to* vs. *too, then* vs. *than*).	0	1	2

Teacher Comments

Here are some things I really liked about your report …	Here are some things I think you could do next time to make your report even better …

Total: _____ /26

Checkout Questions

Investigation 3. Changes in Motion: Where Will the Marble Be Located Each Time It Changes Direction in a Half-Pipe?

The image below shows a half-pipe. Imagine someone places a ball at point A and then lets go of it. Your job is to predict where the ball will be located each time it changes direction as it moves.

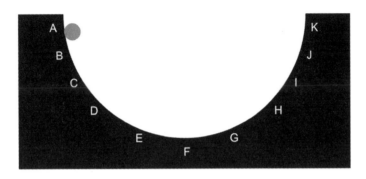

1. Where will the ball be when it changes direction for the first time? H I J K
2. Where will the ball be when it changes direction for the second time? A B C D
3. Where will the ball be when it changes direction for the third time? G H I J
4. Explain your thinking. What pattern from your investigation did you use to decide where the ball would be each time it changed direction as it moved back and forth in the half-pipe?

_____

_____

_____

_____

_____

_____

_____

Teacher Scoring Rubric for the Checkout Questions

Level	Description
3	The student can apply the core idea correctly in all cases and can fully explain the pattern.
2	The student can apply the core idea correctly in all cases but cannot fully explain the pattern.
1	The student cannot apply the core idea correctly in all cases but can fully explain the pattern.
0	The student cannot apply the core idea correctly in all cases and cannot explain the pattern.

Investigation 4

Balanced and Unbalanced Forces: How Do Balanced and Unbalanced Forces Acting on an Object Affect the Motion of That Object?

Introduction

A *force* is a push or a pull. A force can cause an object to move, stop moving, or change how it is moving. Take a moment to explore what happens when you apply more than one pulling force to a cart. You can apply more than one pulling force to a cart by pulling on two strings that are attached to it at the same time. Keep track of what you observe and what you are wondering about in the boxes below.

Things I OBSERVED ...	Things I WONDER about ...

An object will often have more than one force acting on it at the same time. The forces acting on an object can be either balanced or unbalanced. *Balanced forces* are two forces that are the same size but are acting on the object in opposite directions. Scientists describe balanced forces as being equal and opposite. *Unbalanced forces* are not equal and opposite. One example of an unbalanced force is two different-size forces acting on an object in opposite directions. A second example of an unbalanced force is two same-size forces acting on an object in the same direction. When you try to determine if all the forces acting on an object are balanced or unbalanced, it is important to remember that forces that act in the same direction combine by addition and forces that act in opposite directions combine by subtraction.

In this investigation, your goal is to determine how the motion of an object will change when balanced and unbalanced forces act on it. You will be able to observe the relationship between the forces acting an object (a cause) and how the motion of the object changes (the effect) by keeping track of the direction a cart moves when different masses are hung from each side of it. The direction the cart moves will represent the change in motion. The amount of mass hung from each side of the cart will represent the size of the force pulling in opposite directions. You can use this method to examine the relationship between the forces acting on an object and how the motion of the object changes because the two pulling forces will be either balanced or unbalanced, depending on how much mass you decide to hang from each side of the cart.

Things we KNOW from what we read …	What we will NEED to figure out …

Your Task

Use what you know about forces, motion, patterns, and cause and effect to design and carry out an investigation to determine how a cart moves when balanced and unbalanced forces act on it.

The *guiding question* of this investigation is, ***How do balanced and unbalanced forces acting on an object affect the motion of that object?***

Materials

You may use any of the following materials during your investigation:

- Safety glasses or goggles (required)
- Cart
- 2 bench clamp pulleys
- Kite string (6 feet)
- 2 paper clips
- 10 washers

Safety Rules

Follow all normal safety rules. In addition, be sure to follow these rules:

- Wear safety glasses or goggles during setup, investigation activity, and cleanup.
- Do not throw objects or put any objects in your mouth.
- Keep your fingers and toes away from moving objects.
- Wash your hands with soap and water when you are done collecting the data.

Plan Your Investigation

Prepare a plan for your investigation by filling out the chart that follows; this plan is called an *investigation proposal*. Before you start developing your plan, be sure to discuss the following questions with the other members of your group:

- What types of **patterns** might we look for to help answer the guiding question?
- What information do we need to find a **cause-and-effect relationship**?

Our guiding question:

This is a picture of how we will set up the equipment:

We will collect the following data:

These are the steps we will follow to collect data:

I approve of this investigation proposal.

_____
Teacher's signature

_____
Date

Collect Your Data

Keep a record of what you measure or observe during your investigation in the space below.

Analyze Your Data

You will need to analyze the data you collected before you can develop an answer to the guiding question. In the space below, create a table or a graph.

Draft Argument

Develop an argument on a whiteboard. It should include the following parts:

1. A *claim:* Your answer to the guiding question.

2. *Evidence:* An analysis of the data and an explanation of what the analysis means.

3. A *justification of the evidence:* Why your group thinks the evidence is important.

The Guiding Question:	
Our Claim:	
Our Evidence:	Our Justification of the Evidence:

Argumentation Session

Share your argument with your classmates. Be sure to ask them how to make your draft argument better. Keep track of their suggestions in the space below.

Ways to IMPROVE our argument …

Draft Report

Prepare an *investigation report* to share what you have learned. Use the information in this handout and your group's final argument to write a *draft* of your investigation report.

Introduction

We have been studying _____ in class.

Before we started this investigation, we explored _____

_____

_____

We noticed _____

_____

_____

My goal for this investigation was to figure out _____

_____

_____

The guiding question was _____

_____

_____

Method

To gather the data I needed to answer this question, I _____

_____

_____

_____

_____

_____

Investigation Log

I then analyzed the data I collected by _____

_____

_____

_____

Argument

My claim is _____

_____

_____

_____

The _____ below shows _____

_____

_____

_____

_____

_____

_____

National Science Teachers Association

This evidence is important because _____

_____

_____

_____

_____

_____

_____

_____

 **Review**

Your friends need your help! Review the draft of their investigation reports and give them ideas about how to improve. Use the *peer-review guide* that begins on the next page to guide your review.

Investigation Log

Peer-Review Guide

Section 1: The Investigation	Reviewer Rating		
1. Did the author do a good job of explaining what the investigation was about?	☐ No	☐ Almost	☐ Yes
2. Did the author do a good job of making the **guiding question** clear?	☐ No	☐ Almost	☐ Yes
3. Did the author do a good job of describing what he or she did to **collect data?**	☐ No	☐ Almost	☐ Yes
4. Did the author do a good job describing **how** he or she **analyzed** the data?	☐ No	☐ Almost	☐ Yes

Reviewers: If your group gave the author any "No" or "Almost" ratings, please give the author some advice about what to do to improve this part of his or her investigation report.

Section 2: The Argument	Reviewer Rating		
1. Does the author's claim provide a clear and detailed **answer** to the guiding question?	☐ No	☐ Almost	☐ Yes
2. Did the author support his or her claim with **scientific evidence?** Scientific evidence includes analyzed data and an explanation of the analysis.	☐ No	☐ Almost	☐ Yes
3. Does the **evidence** that the author uses in his or her argument **support the claim?**	☐ No	☐ Almost	☐ Yes
4. Did the author include enough **evidence** in his or her argument?	☐ No	☐ Almost	☐ Yes
5. Did the author do a good job of **explaining why the evidence** is important (why it matters)?	☐ No	☐ Almost	☐ Yes
6. Is the content of the argument **correct** based on the science concepts we talked about in class?	☐ No	☐ Almost	☐ Yes

Reviewers: If your group gave the author any "No" or "Almost" ratings, please give the author some advice about what to do to improve this part of his or her investigation report.

Continued

Section 3: Mechanics	Reviewer Rating		
1. *Grammar:* Are the sentences complete? Is there proper subject-verb agreement in each sentence? Are there no run-on sentences?	☐ No	☐ Almost	☐ Yes
2. *Conventions:* Did the author use proper spelling, punctuation, and capitalization?	☐ No	☐ Almost	☐ Yes
3. *Word Choice:* Did the author use the right words in each sentence (for example, *there* vs. *their, to* vs. *too, then* vs. *than*)?	☐ No	☐ Almost	☐ Yes

Reviewers: If your group gave the author any "No" or "Almost" ratings, please give the author some advice about what to do to improve the writing mechanics of his or her investigation report.

General Reviewer Comments

We liked …

We wonder …

Write Your Final Report

Once you have received feedback from your friends about your draft report, create your final investigation report in the space that follows.

Introduction

_____

_____

_____

_____

_____

_____

_____

Method

_____

_____

_____

_____

_____

_____

_____

_____

_____

_____

_____

Argument

_____

_____

_____

_____

_____

_____

_____

_____

_____

_____

_____

_____

Investigation Log

Investigation Report Grading Rubric

Section 1: The Investigation	Missing	Somewhat	Yes
	Score		
1. The author explained what the investigation was about.	0	1	2
2. The author made the **guiding question** clear.	0	1	2
3. The author **described** what he or she did to **collect data.**	0	1	2
4. The author described **how** he or she **analyzed** the data.	0	1	2

Section 2: The Argument	Missing	Somewhat	Yes
	Score		
1. The claim includes a clear and detailed **answer** to the guiding question.	0	1	2
2. The author used **scientific evidence** to support the claim. Scientific evidence includes analyzed data and an explanation of the analysis.	0	1	2
3. The evidence **supports the claim.**	0	1	2
4. The author included enough **evidence** in his or her argument.	0	1	2
5. The author **explained why the evidence** is important.	0	1	2
6. The content of the argument is **correct.**	0	1	2

Section 3: Mechanics	Missing	Somewhat	Yes
	Score		
1. *Grammar:* The sentences are complete. There is proper subject-verb agreement in each sentence. There are no run-on sentences.	0	1	2
2. *Conventions:* The author used proper spelling, punctuation, and capitalization.	0	1	2
3. *Word Choice:* The author used the right words in each sentence (e.g., *there* vs. *their, to* vs. *too, then* vs. *than*).	0	1	2

Teacher Comments

Here are some things I really liked about your report …	Here are some things I think you could do next time to make your report even better …

Total: _____ /26

National Science Teachers Association

Checkout Questions

Investigation 4. Balanced and Unbalanced Forces: How Do Balanced and Unbalanced Forces Acting on an Object Affect the Motion of That Object?

1. Shown below are five different carts. Each cart has a different amount of mass hanging from each side of it. Imagine that you are holding on to each of these carts so they cannot move. Which way will each cart move when you let go of it?

A.

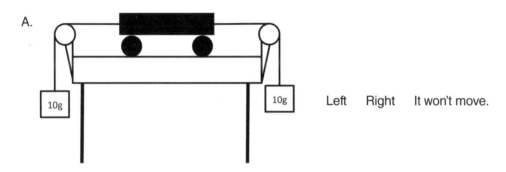

Left   Right   It won't move.

B.

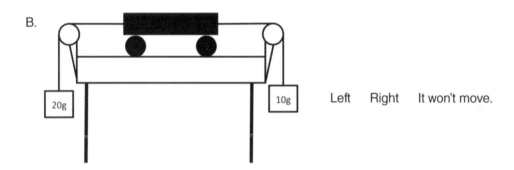

Left   Right   It won't move.

C.

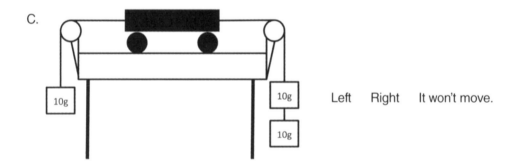

Left   Right   It won't move.

D.

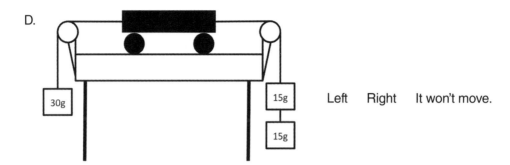

30g   15g

                15g

Left   Right   It won't move.

E.

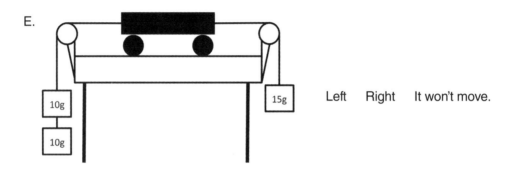

10g   15g

10g

Left   Right   It won't move.

2. Explain your thinking. What *cause-and-effect relationship* did you use to determine which way the cart would move after you let go of it?

_____

_____

_____

_____

_____

_____

_____

_____

Teacher Scoring Rubric for the Checkout Questions

Level	Description
3	The student can apply the core idea correctly in all cases and can fully explain the cause-and-effect relationship.
2	The student can apply the core idea correctly in all cases but cannot fully explain the cause-and-effect relationship.
1	The student cannot apply the core idea correctly in all cases but can fully explain the cause-and-effect relationship.
0	The student cannot apply the core idea correctly in all cases and cannot explain the cause-and-effect relationship.

Section 3
From Molecules to Organisms: Structures and Process

Investigation 5

Life Cycles: How Are the Life Cycles of Living Things Similar and How Are They Different?

Introduction

All living things change over time. A tree starts life as a seed, tadpoles become frogs, and puppies turn into dogs. Take a few minutes to examine how a living thing changes over time as it ages. Keep track of what you observe and what you are wondering about in the boxes below.

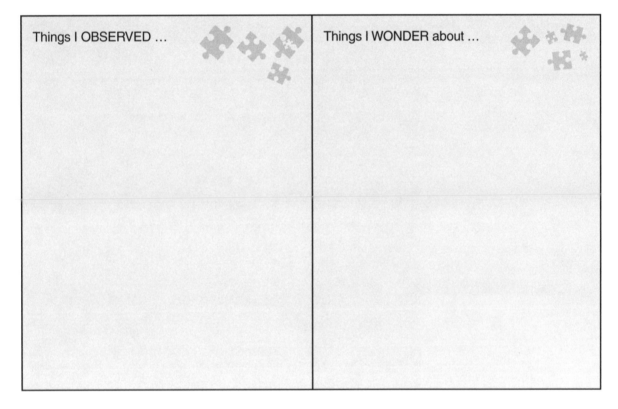

Things I OBSERVED …	Things I WONDER about …

A living thing will follow a predictable pattern of change as it ages. Scientists call this pattern of change a *life cycle*. A life cycle is a series of stages or events that a living thing will go through during its life. Every type of living thing has a specific life cycle.

Think about the life of bird as an example. A bird starts its life within an egg. When a bird hatches from its egg, it is called a hatchling. While a bird lives in a nest and is fed by its parents, it is called a nestling. When a young bird develops its first flight feathers, it is called a fledgling. A young bird

that can fly but is not ready to have offspring (another word for "babies") yet is called a juvenile. A full-grown bird that is ready to have offspring is called an adult. The stage of a bird's life that involves mating and then caring for eggs, hatchlings, and fledglings is called parenthood. After parenthood, a bird will grow old and die. All birds go through these same stages of life. This series of stages of growth and development is the unique life cycle of birds. Other living things, such as butterflies, frogs, and corn, will also follow a predictable pattern of growth and development as they age.

In this investigation you will observe the life cycles of several different kinds of animals and several different kinds of plants. Your goal is to determine what makes the life cycles of these living things similar and what makes them different. To accomplish this goal, you will need to look for patterns in the way different types of living things grow and develop over time. Scientists often look for patterns in nature like this and then use these patterns to classify living things into groups. You can therefore use patterns about how living things grow and develop over time to help determine what is similar about life cycles and what makes them unique.

Things we KNOW from what we read …	What we will NEED to figure out …

Your Task

Use what you know about plants, animals, and patterns to design and carry out an investigation to learn more about the life cycles of different plants and animals.

The *guiding question* of this investigation is, **How are the life cycles of living things similar and how are they different?**

Materials

You may use any of the following materials during your investigation:

- Bean germination in acrylic
- Peanut germination in acrylic
- Life cycle of cabbage worm / cabbage butterfly in acrylic
- Life cycle of dragonfly in acrylic
- Life cycle of fern in acrylic
- Life cycle of frog in acrylic
- Life cycle of grasshopper in acrylic
- Life cycle of silkworm in acrylic

Safety Rules

Follow all normal safety rules. In addition, be sure to follow these rules:

- Do not throw objects or put any objects in your mouth.
- Wash your hands with soap and water when you are done collecting the data.

Plan Your Investigation

Prepare a plan for your investigation by filling out the chart that follows; this plan is called an *investigation proposal*. Before you start developing your plan, be sure to discuss the following questions with the other members of your group:

- What information should we collect so we can **compare and contrast** the life cycles?
- What types of **patterns** might we look for to help answer the guiding question?

Our guiding question:

We will collect the following data:

These are the steps we will follow to collect data:

I approve of this investigation proposal.

_____   _____
Teacher's signature   Date

Collect Your Data

Keep a record of what you measure or observe during your investigation in the space below.

Analyze Your Data

You will need to analyze the data you collected before you can develop an answer to the guiding question. In the space below, you can create a graph that shows how many organisms have the same stage of a life cycle.

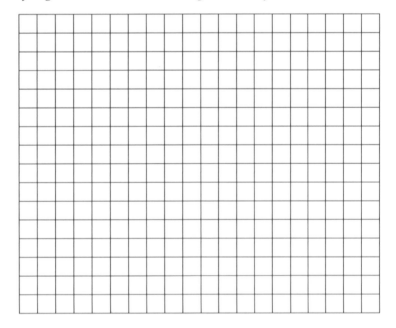

Draft Argument

Develop an argument on a whiteboard. It should include the following parts:

1. A *claim:* Your answer to the guiding question.
2. *Evidence:* An analysis of the data and an explanation of what the analysis means.
3. A *justification of the evidence:* Why your group thinks the evidence is important.

The Guiding Question:	
Our Claim:	
Our Evidence:	Our Justification of the Evidence:

Argumentation Session

Share your argument with your classmates. Be sure to ask them how to make your draft argument better. Keep track of their suggestions in the space below.

Ways to IMPROVE our argument ...

Draft Report

Prepare an *investigation report* to share what you have learned. Use the information in this handout and your group's final argument to write a *draft* of your investigation report.

Introduction

We have been studying _____ in class.

Before we started this investigation, we explored _____

_____

_____

We noticed _____

_____

_____

My goal for this investigation was to figure out _____

_____

_____

_____

The guiding question was _____

_____

_____

_____

Method

To gather the data I needed to answer this question, I _____

_____

_____

_____

_____

Investigation Log

I then analyzed the data I collected by _____

_____

_____

_____

_____

Argument

My claim is _____

_____

_____

The graph below shows _____

_____

_____

_____

_____

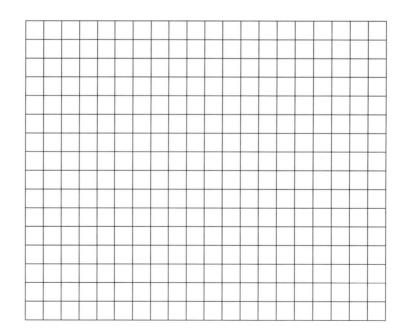

National Science Teachers Association

This evidence is important because _____

_____

_____

_____

_____

_____

_____

_____

_____

 **Review**

Your friends need your help! Review the draft of their investigation reports and give them ideas about how to improve. Use the *peer-review guide* that begins on the next page to guide your review.

Investigation Log

Peer-Review Guide

Section 1: The Investigation	Reviewer Rating		
1. Did the author do a good job of explaining what the investigation was about?	☐ No	☐ Almost	☐ Yes
2. Did the author do a good job of making the **guiding question** clear?	☐ No	☐ Almost	☐ Yes
3. Did the author do a good job of describing what he or she did to **collect data?**	☐ No	☐ Almost	☐ Yes
4. Did the author do a good job describing **how** he or she **analyzed** the data?	☐ No	☐ Almost	☐ Yes

Reviewers: If your group gave the author any "No" or "Almost" ratings, please give the author some advice about what to do to improve this part of his or her investigation report.

Section 2: The Argument	Reviewer Rating		
1. Does the author's claim provide a clear and detailed **answer** to the guiding question?	☐ No	☐ Almost	☐ Yes
2. Did the author support his or her claim with **scientific evidence?** Scientific evidence includes analyzed data and an explanation of the analysis.	☐ No	☐ Almost	☐ Yes
3. Does the **evidence** that the author uses in his or her argument **support the claim?**	☐ No	☐ Almost	☐ Yes
4. Did the author include enough **evidence** in his or her argument?	☐ No	☐ Almost	☐ Yes
5. Did the author do a good job of **explaining why the evidence** is important (why it matters)?	☐ No	☐ Almost	☐ Yes
6. Is the content of the argument **correct** based on the science concepts we talked about in class?	☐ No	☐ Almost	☐ Yes

Reviewers: If your group gave the author any "No" or "Almost" ratings, please give the author some advice about what to do to improve this part of his or her investigation report.

Continued

National Science Teachers Association

Section 3: Mechanics	Reviewer Rating		
1. *Grammar:* Are the sentences complete? Is there proper subject-verb agreement in each sentence? Are there no run-on sentences?	☐ No	☐ Almost	☐ Yes
2. *Conventions:* Did the author use proper spelling, punctuation, and capitalization?	☐ No	☐ Almost	☐ Yes
3. *Word Choice:* Did the author use the right words in each sentence (for example, *there* vs. *their, to* vs. *too, then* vs. *than*)?	☐ No	☐ Almost	☐ Yes

Reviewers: If your group gave the author any "No" or "Almost" ratings, please give the author some advice about what to do to improve the writing mechanics of his or her investigation report.

General Reviewer Comments

We liked …

We wonder …

Write Your Final Report

Once you have received feedback from your friends about your draft report, create your final investigation report in the space that follows.

Introduction

_____

_____

_____

_____

_____

_____

_____

Method

_____

_____

_____

_____

_____

_____

_____

_____

_____

_____

_____

_____

National Science Teachers Association

Argument

_____

_____

_____

_____

_____

_____

_____

_____

_____

_____

_____

_____

Investigation Log

Investigation Report Grading Rubric

Section 1: The Investigation	Score Missing	Somewhat	Yes
1. The author explained what the investigation was about.	0	1	2
2. The author made the **guiding question** clear.	0	1	2
3. The author **described** what he or she did to **collect data.**	0	1	2
4. The author described **how** he or she **analyzed** the data.	0	1	2

Section 2: The Argument	Score Missing	Somewhat	Yes
1. The claim includes a clear and detailed **answer** to the guiding question.	0	1	2
2. The author used **scientific evidence** to support the claim. Scientific evidence includes analyzed data and an explanation of the analysis.	0	1	2
3. The evidence **supports the claim.**	0	1	2
4. The author included enough **evidence** in his or her argument.	0	1	2
5. The author **explained why the evidence** is important.	0	1	2
6. The content of the argument is **correct.**	0	1	2

Section 3: Mechanics	Score Missing	Somewhat	Yes
1. *Grammar:* The sentences are complete. There is proper subject-verb agreement in each sentence. There are no run-on sentences.	0	1	2
2. *Conventions:* The author used proper spelling, punctuation, and capitalization.	0	1	2
3. *Word Choice:* The author used the right words in each sentence (e.g., *there* vs. *their, to* vs. *too, then* vs. *than*).	0	1	2

Teacher Comments

Here are some things I really liked about your report …	Here are some things I think you could do next time to make your report even better …

Total: _____ /26

Checkout Questions

Investigation 5. Life Cycles: How Are the Life Cycles of Living Things Similar and How Are They Different?

1. Listed below are some living things. Place an *X* next to the living things that are born at some point in their life cycle.

    ☐ Corn   ☐ Whale

    ☐ Mouse   ☐ Cactus

    ☐ Tree   ☐ Lizard

2. Listed below are some living things. Place an *X* next to the living things that die at some point in their life cycle.

    ☐ Panda   ☐ Ladybug

    ☐ Snakes ☐ Ants

    ☐ Daisies ☐ Cheetah

3. Listed below are some living things. Place an *X* next to the living things that grow at some point during their life cycle.

    ☐ Spider   ☐ Salmon

    ☐ Bald eagle ☐ Clown fish

    ☐ Grasshopper ☐ Grass

4. Listed below are some living things. Place an *X* next to the living things that reproduce at some point during their life cycle.

    ☐ Bean   ☐ Tree

    ☐ Silkworm ☐ Dragonfly

    ☐ Frog   ☐ Alligator

Checkout Questions

5. Listed below are some living things. Place an *X* next to the living things that go through a complete metamorphosis at some point during their life cycle.

☐ Bean   ☐ Grasshopper

☐ Silkworm   ☐ Dragonfly

☐ Frog   ☐ Butterfly

6. Explain your thinking. What *pattern* from your investigation did you use to decide if these living things are born, grow, reproduce, and die during their life cycle?

_____

_____

_____

_____

_____

_____

_____

_____

_____

Teacher Scoring Rubric for the Checkout Questions

Level	Description
3	The student can apply the core idea correctly in all cases and can fully explain the pattern.
2	The student can apply the core idea correctly in all cases but cannot fully explain the pattern.
1	The student cannot apply the core idea correctly in all cases but can fully explain the pattern.
0	The student cannot apply the core idea correctly in all cases and cannot explain the pattern.

Investigation 6

Life in Groups: Why Do Wolves Live in Groups?

Introduction

All animals must eat to survive. Some animals eat plants, and some animals eat other animals. The musk ox is an example of an animal that eats plants. The arctic wolf is an example of an animal that eats other animals. Both of these animals live in the Arctic tundra. Arctic wolves often eat musk oxen ("oxen" means more than one ox). Take a few minutes to watch what happens when a group of wolves attacks a group of musk oxen. As you watch the video, keep track of what you observe and what you are wondering about in the boxes below.

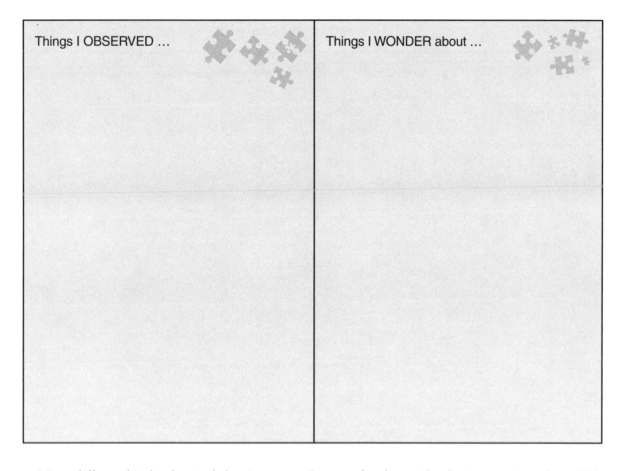

Things I OBSERVED …	Things I WONDER about …

Many different kinds of animals live in groups. Insects often live with other insects in a colony. Fish often travel together in schools. Birds live with other birds in colonies and fly in flocks. Mammals often group together into packs or herds. The size of these groups can range from two or three animals to many thousands of animals.

Wolves are an example of an animal that lives in a group. Scientists often observe 5 to 15 wolves living together for long periods of time. The groups are called wolf packs. There are many potential reasons that may explain why animals, such as wolves, live in a group rather than alone. For example, groups of animals can work together to find food, raise young, or deal with changes in the environment. All of these reasons could make it easier for an animal to survive. Not all animals, however, live in groups. Some animals spend most of their life alone. Therefore, it is important for us to determine why it is a benefit or why it is not a benefit for animals to live as part of a group.

In this investigation you will watch several videos of wolves hunting different types of prey such as caribou, elk, and bison. These three different types of animals are not all the same size. An adult caribou weighs between 200 and 400 pounds, an adult elk weighs between 500 and 700 pounds, and an adult bison weighs between 1,300 and 1,500 pounds. Young caribou, elk, and bison, however, weigh much less.

Your goal in this investigation is to figure out if living in a group (the cause) makes it easier for wolves to get the food they need to survive (the effect). To accomplish this goal, you will need to look for a potential cause-and-effect relationship. Scientists often look for cause-and-effect relationships like this to help explain their observations. You can therefore look for a cause-and-effect relationship to help explain why wolves live in groups.

Things we KNOW from what we read …	What we will NEED to figure out …

Your Task

Use what you know about predators, prey, patterns, and cause-and-effect relationships to design and carry out an investigation to figure out if wolves benefit from hunting in a group.

The *guiding question* of this investigation is, ***Why do wolves live in groups?***

Materials

You will use a computer or tablet with internet access to watch the following videos during your investigation:

- Video showing wolves hunting caribou
- Video showing wolves hunting elk
- Video showing wolves hunting caribou
- Video showing gray wolves chasing down elk
- Video showing baby bison taking on a wolf
- Video showing wolves hunting buffalo
- Video showing wolves taking down elk
- Video showing bison and her calf battling wolves

Safety Rules

Follow all normal safety rules.

Plan Your Investigation

Prepare a plan for your investigation by filling out the chart that follows; this plan is called an *investigation proposal*. Before you start developing your plan, be sure to discuss the following questions with the other members of your group:

- What types of **patterns** might we look for to help answer the guiding question?
- What information do we need to find a **cause-and-effect relationship?**

Investigation Log

Our guiding question:

We will collect the following data from the videos:

These are the steps we will follow to collect data as we watch the videos:

I approve of this investigation proposal.

_____   _____
Teacher's signature   Date

National Science Teachers Association

Collect Your Data

Keep a record of what you observe as you watch the videos in the space below.

Analyze Your Data

You will need to analyze the data you collected while watching the videos before you can develop an answer to the guiding question. In the space below, you can create a table or graph to show the outcomes of the different hunts.

Draft Argument

Develop an argument on a whiteboard. It should include the following parts:

1. A *claim:* Your answer to the guiding question.
2. *Evidence:* An analysis of the data and an explanation of what the analysis means.
3. A *justification of the evidence:* Why your group thinks the evidence is important.

The Guiding Question:	
Our Claim:	
Our Evidence:	Our Justification of the Evidence:

Argumentation Session

Share your argument with your classmates. Be sure to ask them how to make your draft argument better. Keep track of their suggestions in the space below.

Ways to IMPROVE our argument …

Draft Report

Prepare an *investigation report* to share what you have learned. Use the information in this handout and your group's final argument to write a *draft* of your investigation report.

Introduction

We have been studying _____ in class.

Before we started this investigation, we explored _____

_____

_____

We noticed _____

_____

_____

My goal for this investigation was to figure out _____

_____

_____

The guiding question was _____

_____

_____

Method

To gather the data I needed to answer this question, I _____

_____

_____

_____

_____

_____

Investigation Log

I then analyzed the data I collected by _____

_____

_____

_____

_____

Argument

My claim is _____

_____

_____

The _____ below shows _____

_____

_____

_____

This evidence is important because _____

_____

_____

_____

_____

_____

_____

 ### Review

Your friends need your help! Review the draft of their investigation reports and give them ideas about how to improve. Use the *peer-review guide* that begins on the next page to guide your review.

Investigation Log

Peer-Review Guide

Section 1: The Investigation	Reviewer Rating		
1. Did the author do a good job of explaining what the investigation was about?	☐ No	☐ Almost	☐ Yes
2. Did the author do a good job of making the **guiding question** clear?	☐ No	☐ Almost	☐ Yes
3. Did the author do a good job of describing what he or she did to **collect data?**	☐ No	☐ Almost	☐ Yes
4. Did the author do a good job describing **how** he or she **analyzed** the data?	☐ No	☐ Almost	☐ Yes

Reviewers: If your group gave the author any "No" or "Almost" ratings, please give the author some advice about what to do to improve this part of his or her investigation report.

Section 2: The Argument	Reviewer Rating		
1. Does the author's claim provide a clear and detailed **answer** to the guiding question?	☐ No	☐ Almost	☐ Yes
2. Did the author support his or her claim with **scientific evidence?** Scientific evidence includes analyzed data and an explanation of the analysis.	☐ No	☐ Almost	☐ Yes
3. Does the **evidence** that the author uses in his or her argument **support the claim?**	☐ No	☐ Almost	☐ Yes
4. Did the author include enough **evidence** in his or her argument?	☐ No	☐ Almost	☐ Yes
5. Did the author do a good job of **explaining why the evidence** is important (why it matters)?	☐ No	☐ Almost	☐ Yes
6. Is the content of the argument **correct** based on the science concepts we talked about in class?	☐ No	☐ Almost	☐ Yes

Reviewers: If your group gave the author any "No" or "Almost" ratings, please give the author some advice about what to do to improve this part of his or her investigation report.

Continued

National Science Teachers Association

Section 3: Mechanics	Reviewer Rating		
1. *Grammar:* Are the sentences complete? Is there proper subject-verb agreement in each sentence? Are there no run-on sentences?	☐ No	☐ Almost	☐ Yes
2. *Conventions:* Did the author use proper spelling, punctuation, and capitalization?	☐ No	☐ Almost	☐ Yes
3. *Word Choice:* Did the author use the right words in each sentence (for example, *there* vs. *their, to* vs. *too, then* vs. *than*)?	☐ No	☐ Almost	☐ Yes

Reviewers: If your group gave the author any "No" or "Almost" ratings, please give the author some advice about what to do to improve the writing mechanics of his or her investigation report.

General Reviewer Comments

We liked …

We wonder …

Write Your Final Report

Once you have received feedback from your friends about your draft report, create your final investigation report in the space that follows.

Introduction

_____

_____

_____

_____

_____

_____

_____

Method

_____

_____

_____

_____

_____

_____

_____

_____

_____

_____

_____

Argument

_____

_____

_____

_____

_____

_____

_____

_____

_____

_____

_____

_____

Investigation Log

Investigation Report Grading Rubric

Section 1: The Investigation	Score Missing	Score Somewhat	Score Yes
1. The author explained what the investigation was about.	0	1	2
2. The author made the **guiding question** clear.	0	1	2
3. The author **described** what he or she did to **collect data.**	0	1	2
4. The author described **how** he or she **analyzed** the data.	0	1	2

Section 2: The Argument	Score Missing	Score Somewhat	Score Yes
1. The claim includes a clear and detailed **answer** to the guiding question.	0	1	2
2. The author used **scientific evidence** to support the claim. Scientific evidence includes analyzed data and an explanation of the analysis.	0	1	2
3. The evidence **supports the claim.**	0	1	2
4. The author included enough **evidence** in his or her argument.	0	1	2
5. The author **explained why the evidence** is important.	0	1	2
6. The content of the argument is **correct.**	0	1	2

Section 3: Mechanics	Score Missing	Score Somewhat	Score Yes
1. *Grammar:* The sentences are complete. There is proper subject-verb agreement in each sentence. There are no run-on sentences.	0	1	2
2. *Conventions:* The author used proper spelling, punctuation, and capitalization.	0	1	2
3. *Word Choice:* The author used the right words in each sentence (e.g., *there* vs. *their, to* vs. *too, then* vs. *than*).	0	1	2

Teacher Comments

Here are some things I really liked about your report …	Here are some things I think you could do next time to make your report even better …

Total: _____ /26

National Science Teachers Association

Checkout Questions

Investigation 6. Life in Groups: Why Do Wolves Live in Groups?

1. Pictured below are four different animals. Circle the number of wolves that you think would need to hunt together to catch and eat that animal.

A.

Adult moose
1,600–1,800 pounds

1–2   4–6   8–10

B.

Adult caribou
200–400 pounds

1–2   4–6   8–10

C.

Adult white-tailed deer
80–100 pounds

1–2   4–6   8–10

D.

Baby moose
50–80 pounds

1–2   4–6   8–10

Checkout Questions

2. Explain your thinking. What *cause-and-effect relationship* did you use to determine how many wolves would need to hunt together to catch and eat an animal?

_____

_____

_____

_____

_____

_____

_____

_____

_____

Teacher Scoring Rubric for the Checkout Questions

Level	Description
3	The student can apply the core idea correctly in all cases and can fully explain the cause-and-effect relationship.
2	The student can apply the core idea correctly in all cases but cannot fully explain the cause-and-effect relationship.
1	The student cannot apply the core idea correctly in all cases but can fully explain the cause-and-effect relationship.
0	The student cannot apply the core idea correctly in all cases and cannot explain the cause-and-effect relationship.

Section 4
Heredity: Inheritance and Variation of Traits

Investigation 7

Variation Within a Species: How Similar Are Earthworms to Each Other?

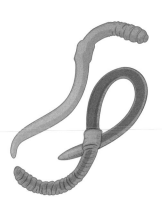

Introduction

When you look at the animals living outside your home or school, you will probably notice a few things. First, there are many different types of animals. Second, you might notice that animals that are the same type share a lot of traits. For example, squirrels have bushy tails, brown eyes, and four legs. Robins have wings, beaks, and two legs. But animals that are the same type of animal can also have some traits that make them look different. For example, not all squirrels are the same size or have the same color fur.

Your teacher will give you two earthworms. Take a moment to observe these two earthworms. Keep track of what you observe and what you are wondering about in the table below.

Things I OBSERVED …	Things I WONDER about …

Scientists classify animals based on the traits they share. For example, dogs have two eyes, four legs, four toes that touch the ground on each foot, a tail, and hair, and they can bark. When an animal has these traits, we call it a dog. There will also be differences in the traits that the animals that belong to a specific group share, because traits come in different versions. For example, an animal can have a tail but the tail can be long or short. The different lengths of tail are different versions of the same trait. No two animals, as a result, will look exactly alike even when they are members of the same group. For example, all dogs have two eyes but some dogs have dark brown eyes and some have blue eyes. All dogs have hair, but the hair can be different colors (such as black, brown, or yellow). Therefore, no two dogs look exactly alike because they each have different versions of the traits that all dogs share.

Your goal in this investigation is to figure out how similar earthworms are to each other. To accomplish this task, you will need to make observations about and take measurements of the traits of earthworms. You will need to compare and contrast at least two different traits to figure out what the worms have in common and what is different about them. Scientists often look for patterns in nature like this and then use these patterns to classify animals into groups. You can therefore use patterns to help determine what is similar about earthworms that make them all the same species of animal, and what is different about individual worms even though they are the same species.

Things we KNOW from what we read …	What we will NEED to figure out …

National Science Teachers Association

Your Task

Use what you know about traits and patterns to design and carry out an investigation to compare and contrast at least two different traits of earthworms.

The *guiding question* of this investigation is, ***How similar are earthworms to each other?***

Materials

You may use any of the following materials during your investigation:

- Safety goggles (required)
- Nitrile gloves (required)
- Earthworms
- Ruler
- Electronic scale
- Magnifying glass
- Tray

Safety Rules

Follow all normal safety rules. In addition, be sure to keep the earthworms healthy by not doing anything to hurt them such as pulling on them, poking them, cutting them, or dropping them. You also need to follow these rules:

- Wear sanitized indirectly vented chemical-splash goggles and nitrile gloves during setup, investigation activity, and cleanup.
- Wash your hands with soap and water when you are done collecting the data.

Plan Your Investigation

Prepare a plan for your investigation by filling out the chart that follows; this plan is called an *investigation proposal*. Before you start developing your plan, be sure to discuss the following questions with the other members of your group:

- What information should we collect so we can **describe** the traits of an earthworm?
- What types of **patterns** might we look for to help answer the guiding question?

Our guiding question:

We will collect the following data:

These are the steps we will follow to collect data:

I approve of this investigation proposal.

_____
Teacher's signature

_____
Date

Collect Your Data

Keep a record of what you measure or observe during your investigation in the space below.

Analyze Your Data

You will need to analyze the data you collected before you can develop an answer to the guiding question. To do this, create one or more graphs that show how many earthworms had a specific version of a trait.

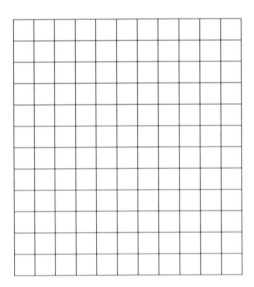

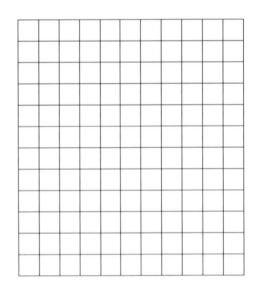

Draft Argument

Develop an argument on a whiteboard. It should include the following parts:

1. A *claim:* Your answer to the guiding question.
2. *Evidence:* An analysis of the data and an explanation of what the analysis means.
3. A *justification of the evidence:* Why your group thinks the evidence is important.

The Guiding Question:	
Our Claim:	
Our Evidence:	Our Justification of the Evidence:

Argumentation Session

Share your argument with your classmates. Be sure to ask them how to make your draft argument better. Keep track of their suggestions in the space below.

Ways to IMPROVE our argument …

National Science Teachers Association

Draft Report

Prepare an *investigation report* to share what you have learned. Use the information in this handout and your group's final argument to write a *draft* of your investigation report.

Introduction

We have been studying _____ in class.

Before we started this investigation, we explored _____

_____

_____

We noticed _____

_____

_____

My goal for this investigation was to figure out _____

_____

_____

_____

The guiding question was _____

_____

_____

Method

To gather the data I needed to answer this question, I _____

_____

_____

_____

_____

_____

Investigation Log

I then analyzed the data I collected by _____

_____

_____

_____

Argument

My claim is _____

_____

_____

_____

The graphs below include information about _____

_____

_____

This analysis of the data I collected shows _____

_____

_____

_____

This evidence is important because _____

_____

_____

_____

_____

_____

_____

_____

 **Review**

Your friends need your help! Review the draft of their investigation reports and give them ideas about how to improve. Use the *peer-review guide* that begins on the next page to guide your review.

Investigation Log

Peer-Review Guide

Section 1: The Investigation	Reviewer Rating		
1. Did the author do a good job of explaining what the investigation was about?	☐ No	☐ Almost	☐ Yes
2. Did the author do a good job of making the **guiding question** clear?	☐ No	☐ Almost	☐ Yes
3. Did the author do a good job of describing what he or she did to **collect data?**	☐ No	☐ Almost	☐ Yes
4. Did the author do a good job describing **how** he or she **analyzed** the data?	☐ No	☐ Almost	☐ Yes

Reviewers: If your group gave the author any "No" or "Almost" ratings, please give the author some advice about what to do to improve this part of his or her investigation report.

Section 2: The Argument	Reviewer Rating		
1. Does the author's claim provide a clear and detailed **answer** to the guiding question?	☐ No	☐ Almost	☐ Yes
2. Did the author support his or her claim with **scientific evidence?** Scientific evidence includes analyzed data and an explanation of the analysis.	☐ No	☐ Almost	☐ Yes
3. Does the **evidence** that the author uses in his or her argument **support the claim?**	☐ No	☐ Almost	☐ Yes
4. Did the author include enough **evidence** in his or her argument?	☐ No	☐ Almost	☐ Yes
5. Did the author do a good job of **explaining why the evidence** is important (why it matters)?	☐ No	☐ Almost	☐ Yes
6. Is the content of the argument **correct** based on the science concepts we talked about in class?	☐ No	☐ Almost	☐ Yes

Reviewers: If your group gave the author any "No" or "Almost" ratings, please give the author some advice about what to do to improve this part of his or her investigation report.

Continued

Section 3: Mechanics	Reviewer Rating		
1. *Grammar:* Are the sentences complete? Is there proper subject-verb agreement in each sentence? Are there no run-on sentences?	☐ No	☐ Almost	☐ Yes
2. *Conventions:* Did the author use proper spelling, punctuation, and capitalization?	☐ No	☐ Almost	☐ Yes
3. *Word Choice:* Did the author use the right words in each sentence (for example, *there* vs. *their, to* vs. *too, then* vs. *than*)?	☐ No	☐ Almost	☐ Yes

Reviewers: If your group gave the author any "No" or "Almost" ratings, please give the author some advice about what to do to improve the writing mechanics of his or her investigation report.

General Reviewer Comments

We liked …

We wonder …

Write Your Final Report

Once you have received feedback from your friends about your draft report, create your final investigation report in the space that follows.

Introduction

_____

_____

_____

_____

_____

_____

_____

Method

_____

_____

_____

_____

_____

_____

_____

_____

_____

_____

Argument

_____

_____

_____

_____

_____

_____

_____

_____

_____

_____

_____

_____

_____

Investigation Log

Investigation Report Grading Rubric

Section 1: The Investigation	Score		
	Missing	Somewhat	Yes
1. The author explained what the investigation was about.	0	1	2
2. The author made the **guiding question** clear.	0	1	2
3. The author **described** what he or she did to **collect data.**	0	1	2
4. The author described **how** he or she **analyzed** the data.	0	1	2

Section 2: The Argument	Score		
	Missing	Somewhat	Yes
1. The claim includes a clear and detailed **answer** to the guiding question.	0	1	2
2. The author used **scientific evidence** to support the claim. Scientific evidence includes analyzed data and an explanation of the analysis.	0	1	2
3. The evidence **supports the claim.**	0	1	2
4. The author included enough **evidence** in his or her argument.	0	1	2
5. The author **explained why the evidence** is important.	0	1	2
6. The content of the argument is **correct.**	0	1	2

Section 3: Mechanics	Score		
	Missing	Somewhat	Yes
1. *Grammar:* The sentences are complete. There is proper subject-verb agreement in each sentence. There are no run-on sentences.	0	1	2
2. *Conventions:* The author used proper spelling, punctuation, and capitalization.	0	1	2
3. *Word Choice:* The author used the right words in each sentence (e.g., *there* vs. *their, to* vs. *too, then* vs. *than*).	0	1	2

Teacher Comments

Here are some things I really liked about your report …	Here are some things I think you could do next time to make your report even better …

Total: _____ /26

Checkout Questions

Investigation 7. Variation Within a Species: How Similar Are Earthworms to Each Other?

1. Pictured below is an earthworm. Place an "S" next to the traits that you think will be the same when you compare two different earthworms and a "D" next to the traits that you think will be different.

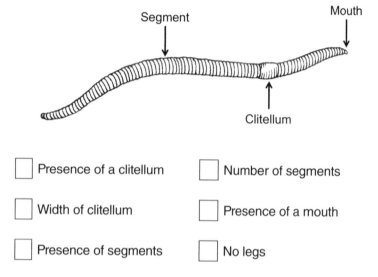

Segment   Mouth   Clitellum

☐ Presence of a clitellum   ☐ Number of segments

☐ Width of clitellum   ☐ Presence of a mouth

☐ Presence of segments   ☐ No legs

2. Pictured below is a ladybug. Place an "S" next to the traits that you think will be the same when you compare two different ladybugs and a "D" next to the traits that you think will be different.

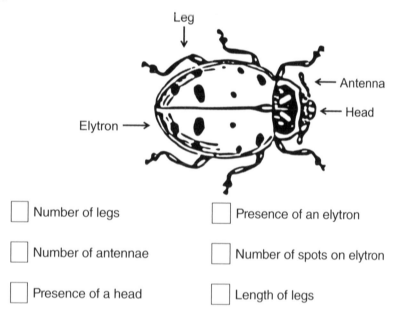

Leg   Antenna   Head   Elytron

☐ Number of legs   ☐ Presence of an elytron

☐ Number of antennae   ☐ Number of spots on elytron

☐ Presence of a head   ☐ Length of legs

Checkout Questions

3. Explain your thinking. What *pattern* from your investigation did you use to predict if a trait would be the same or different when comparing the traits of two individuals that belong to the same group of organisms?

_____

_____

_____

_____

_____

_____

_____

Teacher Scoring Rubric for the Checkout Questions

Level	Description
3	The student can apply the core idea correctly in all cases and can fully explain the pattern.
2	The student can apply the core idea correctly in all cases but cannot fully explain the pattern.
1	The student cannot apply the core idea correctly in all cases but can fully explain the pattern.
0	The student cannot apply the core idea correctly in all cases and cannot explain the pattern.

Investigation 8

Inheritance of Traits: How Similar Are Offspring to Their Parents?

Introduction

All living things have some traits in common with other living things and some traits that make them unique. Your teacher will give you some pictures of dogs. Take a minute to look at the pictures. Be sure to write down what you observe and what you are wondering about in the boxes below.

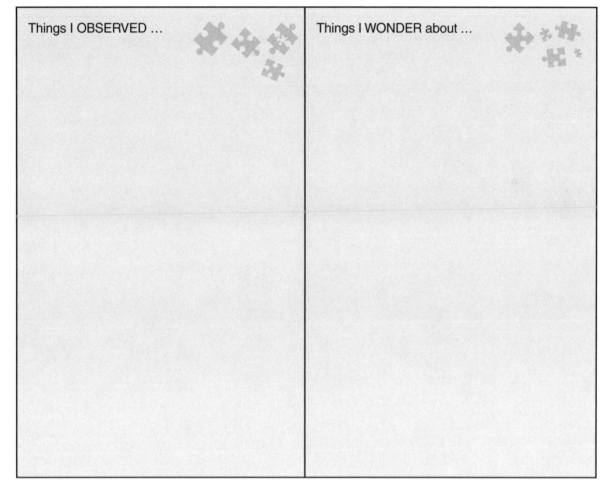

Things I OBSERVED ...	Things I WONDER about ...

The members of any group of living things, such as the group of dogs you just looked at, will often include some individuals that are related to each other and other individuals that are not related. The individuals in the group that are related to each other will look more alike than the individuals in the

group that are not related because animals and plants inherit their traits from their parents. Animals or plants that share the same parents will therefore have many traits in common, and animals or plants that do not share the same parents will have fewer traits in common. You have already seen how some dog traits are the same and others are different. The dogs you examined, however, did not have the same parents. You will now have a chance to examine the traits of parents and children (what scientists call *offspring*).

In this investigation you need to figure out how similar offspring are to their parents. To accomplish this task, you will need to make observations about the traits of parents and offspring for at least two different types of living things. You will need to compare and contrast these traits to figure out which members of the same family share a trait and which members of the same family do not share a trait. You can look for traits that one parent and one offspring have in common; traits that are shared by both parents and an offspring; and traits that are only observed in one or more of the offspring. Your goal is to look for any patterns in the traits that the parents and offspring have in common. Scientists often look for patterns in nature like this and then use these patterns to determine how plants and animals inherit traits.

Things we KNOW from what we read …	What we will NEED to figure out …

Your Task

Use what you know about traits and patterns to design and carry out an investigation to determine which traits parents and offspring have in common and which ones are different.

The *guiding question* of this investigation is, ***How similar are offspring to their parents?***

Materials

You will use at least two of the following sets of pictures of parents and offspring during your investigation:

- Dogs A
- Dogs B
- Cats A
- Cats B
- Guinea pigs A
- Guinea pigs B
- Birds A
- Birds B
- Snapdragons
- Peas
- Tulips

Safety Rules

Follow all normal safety rules.

Plan Your Investigation

Prepare a plan for your investigation by filling out the chart that follows; this plan is called an *investigation proposal*. Before you start developing your plan, be sure to discuss the following questions with the other members of your group:

- What information should we collect so we can **describe** the traits of an animal?
- What types of **patterns** might we look for to help answer the guiding question?

Investigation Log

Our guiding question:

We will collect the following data:

These are the steps we will follow to collect data:

I approve of this investigation proposal.

_____   _____
Teacher's signature   Date

Collect Your Data

Keep a record of what you measure or observe during your investigation in the space below.

Analyze Your Data

You will need to analyze the data you collected before you can develop an answer to the guiding question. In the space below, create a table. Your table should include the names of the animals or plants you observed and the traits that the different individuals did or did not have.

Draft Argument

Develop an argument on a whiteboard. It should include the following parts:

The Guiding Question:	
Our Claim:	
Our Evidence:	Our Justification of the Evidence:

1. A *claim:* Your answer to the guiding question.

2. *Evidence:* An analysis of the data and an explanation of what the analysis means.

3. A *justification of the evidence:* Why your group thinks the evidence is important.

Argumentation Session

Share your argument with your classmates. Be sure to ask them how to make your draft argument better. Keep track of their suggestions in the space below.

Ways to IMPROVE our argument …

Draft Report

Prepare an *investigation report* to share what you have learned. Use the information in this handout and your group's final argument to write a draft of your investigation report.

Introduction

We have been studying _____ in class.

Before we started this investigation, we explored _____

_____

_____

We noticed _____

_____

_____

My goal for this investigation was to figure out _____

_____

_____

_____

The guiding question was _____

_____

_____

Method

To gather the data I needed to answer this question, I _____

_____

_____

_____

_____

_____

Investigation Log

I then analyzed the data I collected by _____

_____

_____

Argument

My claim is _____

_____

_____

_____

The table below shows _____

_____

_____

_____

_____

National Science Teachers Association

This evidence is important because _____

_____

_____

_____

_____

_____

_____

_____

_____

_____

 **Review**

Your friends need your help! Review the draft of their investigation reports and give them ideas about how to improve. Use the *peer-review guide* that begins on the next page to guide your review.

Investigation Log

Peer-Review Guide

Section 1: The Investigation	Reviewer Rating		
1. Did the author do a good job of explaining what the investigation was about?	☐ No	☐ Almost	☐ Yes
2. Did the author do a good job of making the **guiding question** clear?	☐ No	☐ Almost	☐ Yes
3. Did the author do a good job of describing what he or she did to **collect data?**	☐ No	☐ Almost	☐ Yes
4. Did the author do a good job describing **how** he or she **analyzed** the data?	☐ No	☐ Almost	☐ Yes

Reviewers: If your group gave the author any "No" or "Almost" ratings, please give the author some advice about what to do to improve this part of his or her investigation report.

Section 2: The Argument	Reviewer Rating		
1. Does the author's claim provide a clear and detailed **answer** to the guiding question?	☐ No	☐ Almost	☐ Yes
2. Did the author support his or her claim with **scientific evidence?** Scientific evidence includes analyzed data and an explanation of the analysis.	☐ No	☐ Almost	☐ Yes
3. Does the **evidence** that the author uses in his or her argument **support the claim?**	☐ No	☐ Almost	☐ Yes
4. Did the author include enough **evidence** in his or her argument?	☐ No	☐ Almost	☐ Yes
5. Did the author do a good job of **explaining why the evidence** is important (why it matters)?	☐ No	☐ Almost	☐ Yes
6. Is the content of the argument **correct** based on the science concepts we talked about in class?	☐ No	☐ Almost	☐ Yes

Reviewers: If your group gave the author any "No" or "Almost" ratings, please give the author some advice about what to do to improve this part of his or her investigation report.

Continued

Section 3: Mechanics	Reviewer Rating		
1. *Grammar:* Are the sentences complete? Is there proper subject-verb agreement in each sentence? Are there no run-on sentences?	☐ No	☐ Almost	☐ Yes
2. *Conventions:* Did the author use proper spelling, punctuation, and capitalization?	☐ No	☐ Almost	☐ Yes
3. *Word Choice:* Did the author use the right words in each sentence (for example, *there* vs. *their, to* vs. *too, then* vs. *than*)?	☐ No	☐ Almost	☐ Yes

Reviewers: If your group gave the author any "No" or "Almost" ratings, please give the author some advice about what to do to improve the writing mechanics of his or her investigation report.

General Reviewer Comments

We liked …

We wonder …

Write Your Final Report

Once you have received feedback from your friends about your draft report, create your final investigation report in the space that follows.

Introduction

_____

_____

_____

_____

_____

_____

_____

_____

Method

_____

_____

_____

_____

_____

_____

_____

_____

_____

_____

_____

_____

Argument

_____

_____

_____

_____

_____

_____

_____

_____

_____

_____

_____

_____

_____

Investigation Log

Investigation Report Grading Rubric

Section 1: The Investigation	Score		
	Missing	**Somewhat**	**Yes**
1. The author explained what the investigation was about.	0	1	2
2. The author made the **guiding question** clear.	0	1	2
3. The author **described** what he or she did to **collect data.**	0	1	2
4. The author described **how** he or she **analyzed** the data.	0	1	2

Section 2: The Argument	Score		
	Missing	**Somewhat**	**Yes**
1. The claim includes a clear and detailed **answer** to the guiding question.	0	1	2
2. The author used **scientific evidence** to support the claim. Scientific evidence includes analyzed data and an explanation of the analysis.	0	1	2
3. The evidence **supports the claim.**	0	1	2
4. The author included enough **evidence** in his or her argument.	0	1	2
5. The author **explained why the evidence** is important.	0	1	2
6. The content of the argument is **correct.**	0	1	2

Section 3: Mechanics	Score		
	Missing	**Somewhat**	**Yes**
1. *Grammar:* The sentences are complete. There is proper subject-verb agreement in each sentence. There are no run-on sentences.	0	1	2
2. *Conventions:* The author used proper spelling, punctuation, and capitalization.	0	1	2
3. *Word Choice:* The author used the right words in each sentence (e.g., *there* vs. *their, to* vs. *too, then* vs. *than*).	0	1	2

Teacher Comments

Here are some things I really liked about your report …	Here are some things I think you could do next time to make your report even better …

Total: _____ /26

Investigation 8. Inheritance of Traits: How Similar Are Offspring to Their Parents?

1. Pictured below are a female guinea pig and three of her pups.

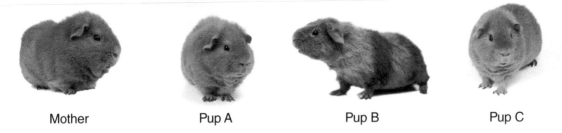

| Mother | Pup A | Pup B | Pup C |

Which of the following guinea pigs is most likely the father of the three pups in this litter?

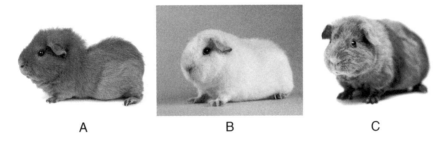

| A | B | C |

Full-color versions of these images are available on the teacher book's Extras page at *www.nsta.org/adi-3rd.*

2. Explain your thinking. What pattern from your investigation did you use to predict the father of the three guinea pig pups?

_____

_____

_____

_____

_____

_____

Teacher Scoring Rubric for the Checkout Questions

Level	Description
3	The student can apply the core idea correctly in all cases and can fully explain the pattern.
2	The student can apply the core idea correctly in all cases but cannot fully explain the pattern.
1	The student cannot apply the core idea correctly in all cases but can fully explain the pattern.
0	The student cannot apply the core idea correctly in all cases and cannot explain the pattern.

Investigation 9

Traits and the Environment: How Do Differences in Soil Quality Affect the Traits of a Plant?

Introduction

Plants have many unique traits that animals do not have. These traits enable plants to turn carbon dioxide into sugar, get water from the soil, and reproduce. Take a moment to examine a plant. Be sure to keep track of what you observe and what you are wondering about in the boxes below.

Things I OBSERVED ...	Things I WONDER about ...

The plant you observed is called a flowering plant. Flowering plants have several leaves, a stem, and one or more flowers. The leaves of plants are very important. Plants use carbon dioxide from the air and water from the soil to make sugar inside their leaves though a process called *photosynthesis*. Plants then use this sugar as a source of energy. Plants are able to move water from the soil to the leaves inside their stems. The flowers are used in reproduction.

Flowering plants produce pollen inside the flowers. This pollen is then spread to other plants by the wind or by sticking to animals such as bees and birds. Once a plant is fertilized by pollen, it will produce seeds that can grow and develop into a new plant.

The leaves, stem, and flowers of flowering plants make them different from other living things, such as animals or fungi.

Flowering plants can have very different-looking leaves, stems, and flowers. The leaves and stems of a plant can be big and wide or small and narrow. Flowers can also come in a wide range of shapes, sizes, and colors. The size, shape, and color of leaves, stems, and flowers are all traits that are passed down from parent to offspring. This is one reason that there are so many different kinds of plants. Adult plants produce offspring with traits that are similar to the ones they have. The environment, however, can also change the traits of a plant as it grows and develops.

In this investigation, your goal is to figure out how a change in a characteristic of soil (a cause) affects one or more different plant traits (the effect). You will be able to change one of two different characteristics of soil during your investigation: either the moisture of the soil or the amount of minerals in the soil. These minerals include the elements of nitrogen, phosphorus, and potassium. You can change the amount of minerals found in the soil by adding one or more fertilizer pellets to it. Plants use water to produce sugar for energy and minerals to create more plant parts (such as leaves and roots).

There are many different plant traits that you can study during your investigation. You can look at the height of the plant, leaf size, the number of leaves, or the color of the leaves. You can also examine a combination of these four different plant traits.

Things we KNOW from what we read …	What we will NEED to figure out …

National Science Teachers Association

Your Task

Use what you know about cause and effect to plan and carry out an investigation to determine how a change in amount of water or minerals in the soil will or will not change the traits of a plant. Be sure to set up a fair test so you can determine if there is a cause-and-effect relationship or not. To do that, you will need to grow the same type of plant under at least three different conditions.

The *guiding question* of this investigation is, ***How do differences in soil quality affect the traits of a plant?***

Materials

You may use any of the following materials during your investigation:

Equipment

- Windowsill greenhouse
- Plant labels
- Ruler
- Plastic pipette

Consumables

- Potting pellets
- Seeds
- Fertilizer pellets

Safety Rules

Follow all normal safety rules. In addition, be sure to follow these rules:

- Wear sanitized indirectly vented chemical-splash goggles during setup, investigation activity, and cleanup.
- Never eat food or drink during the activity using fertilizer pellets.
- Wash your hands with soap and water when you are done collecting the data.

Plan Your Investigation

Prepare a plan for your investigation by filling out the chart that follows; this plan is called an *investigation proposal.* Before you start developing your plan, be sure to discuss the following questions with the other members of your group:

- What information should we collect so we can **describe** the traits of a plant?
- What information do we need to find a relationship between **a cause and an effect?**

Investigation Log

Our guiding question:

This is a picture of how we will set up the equipment:

We will collect the following data:

These are the steps we will follow to collect data:

I approve of this investigation proposal.

_____
Teacher's signature

_____
Date

Collect Your Data

Keep a record of what you measure or observe during your investigation in the space below.

Analyze Your Data

You will need to analyze the data you collected before you can develop an answer to the guiding question. To do this, create a graph that shows the relationship between the cause (change in soil quality) and the effect (traits of the plant).

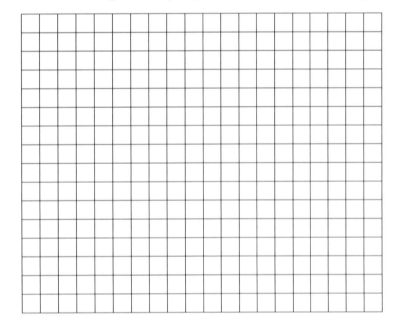

Investigation Log

Draft Argument

Develop an argument on a whiteboard. It should include the following parts:

1. A *claim:* Your answer to the guiding question.

2. *Evidence:* An analysis of the data and an explanation of what the analysis means.

3. A *justification of the evidence:* Why your group thinks the evidence is important.

The Guiding Question:	
Our Claim:	
Our Evidence:	Our Justification of the Evidence:

Argumentation Session

Share your argument with your classmates. Be sure to ask them how to make your draft argument better. Keep track of their suggestions in the space below.

Ways to IMPROVE our argument …

National Science Teachers Association

Draft Report

Prepare an *investigation report* to share what you have learned. Use the information in this handout and your group's final argument to write a *draft* of your investigation report.

Introduction

We have been studying _____ in class.

Before we started this investigation, we explored _____

_____

_____

_____

We noticed _____

_____

_____

My goal for this investigation was to figure out _____

_____

_____

The guiding question was _____

_____

_____

Method

To gather the data I needed to answer this question, I _____

_____

_____

_____

_____

Investigation Log

I then analyzed the data I collected by _____

_____

_____

Argument

My claim is _____

_____

_____

_____

The graph below shows _____

_____

_____

_____

_____

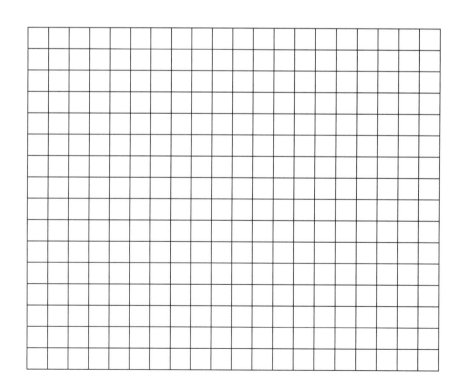

This evidence is important because _____

_____

_____

_____

_____

_____

_____

_____

 **Review**

Your friends need your help! Review the draft of their investigation reports and give them ideas about how to improve. Use the *peer-review guide* that begins on the next page to guide your review.

Investigation Log

Peer-Review Guide

Section 1: The Investigation	Reviewer Rating		
1. Did the author do a good job of explaining what the investigation was about?	☐ No	☐ Almost	☐ Yes
2. Did the author do a good job of making the **guiding question** clear?	☐ No	☐ Almost	☐ Yes
3. Did the author do a good job of describing what he or she did to **collect data?**	☐ No	☐ Almost	☐ Yes
4. Did the author do a good job describing **how** he or she **analyzed** the data?	☐ No	☐ Almost	☐ Yes

Reviewers: If your group gave the author any "No" or "Almost" ratings, please give the author some advice about what to do to improve this part of his or her investigation report.

Section 2: The Argument	Reviewer Rating		
1. Does the author's claim provide a clear and detailed **answer** to the guiding question?	☐ No	☐ Almost	☐ Yes
2. Did the author support his or her claim with **scientific evidence?** Scientific evidence includes analyzed data and an explanation of the analysis.	☐ No	☐ Almost	☐ Yes
3. Does the **evidence** that the author uses in his or her argument **support the claim?**	☐ No	☐ Almost	☐ Yes
4. Did the author include enough **evidence** in his or her argument?	☐ No	☐ Almost	☐ Yes
5. Did the author do a good job of **explaining why the evidence** is important (why it matters)?	☐ No	☐ Almost	☐ Yes
6. Is the content of the argument **correct** based on the science concepts we talked about in class?	☐ No	☐ Almost	☐ Yes

Reviewers: If your group gave the author any "No" or "Almost" ratings, please give the author some advice about what to do to improve this part of his or her investigation report.

Continued

National Science Teachers Association

Section 3: Mechanics	Reviewer Rating		
1. *Grammar:* Are the sentences complete? Is there proper subject-verb agreement in each sentence? Are there no run-on sentences?	☐ No	☐ Almost	☐ Yes
2. *Conventions:* Did the author use proper spelling, punctuation, and capitalization?	☐ No	☐ Almost	☐ Yes
3. *Word Choice:* Did the author use the right words in each sentence (for example, *there* vs. *their, to* vs. *too, then* vs. *than*)?	☐ No	☐ Almost	☐ Yes

Reviewers: If your group gave the author any "No" or "Almost" ratings, please give the author some advice about what to do to improve the writing mechanics of his or her investigation report.

General Reviewer Comments

We liked …

We wonder …

Write Your Final Report

Once you have received feedback from your friends about your draft report, create your final investigation report in the space that follows.

Introduction

_____

_____

_____

_____

_____

_____

_____

Method

_____

_____

_____

_____

_____

_____

_____

_____

_____

_____

Argument

_____

_____

_____

_____

_____

_____

_____

_____

_____

_____

_____

_____

_____

_____

Investigation Log

Investigation Report Grading Rubric

Section 1: The Investigation	Score Missing	Score Somewhat	Score Yes
1. The author explained what the investigation was about.	0	1	2
2. The author made the **guiding question** clear.	0	1	2
3. The author **described** what he or she did to **collect data.**	0	1	2
4. The author described **how** he or she **analyzed** the data.	0	1	2

Section 2: The Argument	Score Missing	Score Somewhat	Score Yes
1. The claim includes a clear and detailed **answer** to the guiding question.	0	1	2
2. The author used **scientific evidence** to support the claim. Scientific evidence includes analyzed data and an explanation of the analysis.	0	1	2
3. The evidence **supports the claim.**	0	1	2
4. The author included enough **evidence** in his or her argument.	0	1	2
5. The author **explained why the evidence** is important.	0	1	2
6. The content of the argument is **correct.**	0	1	2

Section 3: Mechanics	Score Missing	Score Somewhat	Score Yes
1. *Grammar:* The sentences are complete. There is proper subject-verb agreement in each sentence. There are no run-on sentences.	0	1	2
2. *Conventions:* The author used proper spelling, punctuation, and capitalization.	0	1	2
3. *Word Choice:* The author used the right words in each sentence (e.g., *there* vs. *their, to* vs. *too, then* vs. *than*).	0	1	2

Teacher Comments

Here are some things I really liked about your report …	Here are some things I think you could do next time to make your report even better …

Total: _____ /26

Checkout Questions

Investigation 9. Traits and the Environment: How Do Differences in Soil Quality Affect the Traits of a Plant?

Imagine that you plant five seeds in some potting soil. All the seeds are from the same type of plant. Each day you add 2 ml of water to the soil. After a few weeks you have five plants.

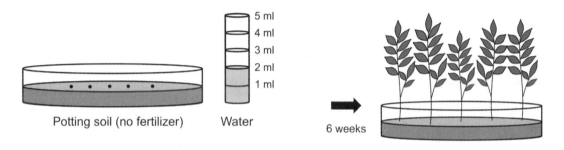

Potting soil (no fertilizer)   Water   6 weeks

1. What would the plants look like if you decided to add fertilizer to the soil? Circle the letter (A, B, or C) under the picture that matches what they would look like.

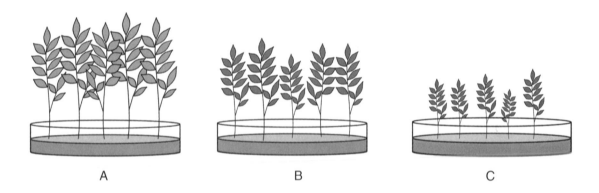

A   B   C

2. What would the plants look like if you decided to add less water to the soil? Circle the letter (A, B, or C) under the picture that matches what they would look like.

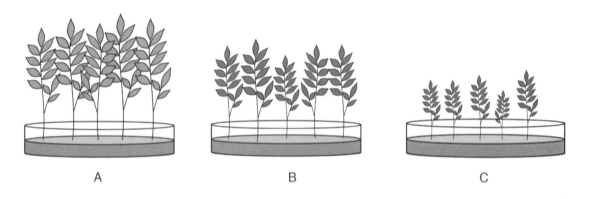

A   B   C

Checkout Questions

3. What would the plants look like if you decided to add more water to the soil? Circle the letter (A, B, or C) under the picture that matches what they would look like.

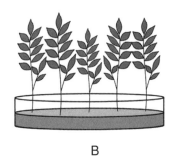
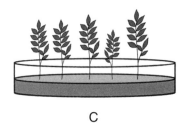

A   B   C

4. Explain your thinking. What *cause-and-effect relationship* did you use to determine how the plants would look under each condition?

_____

_____

_____

_____

_____

_____

_____

_____

_____

_____

_____

Teacher Scoring Rubric for the Checkout Questions

Level	Description
3	The student can apply the core idea correctly in all cases and can fully explain the cause-and-effect relationship.
2	The student can apply the core idea correctly in all cases but cannot fully explain the cause-and-effect relationship.
1	The student cannot apply the core idea correctly in all cases but can fully explain the cause-and-effect relationship.
0	The student cannot apply the core idea correctly in all cases and cannot explain the cause-and-effect relationship.

Section 5

Biological Evolution: Unity and Diversity

Investigation Log

Investigation 10

Fossils: What Was the Ecosystem at Darmstadt Like 49 Million Years Ago?

Introduction

A *fossil* is the preserved remains of a plant or animal. Fossils provide information about organisms that lived long ago and the environments where these organisms lived. Take a minute to examine a fossil. Keep track of what you observe and what you are wondering about in the boxes below.

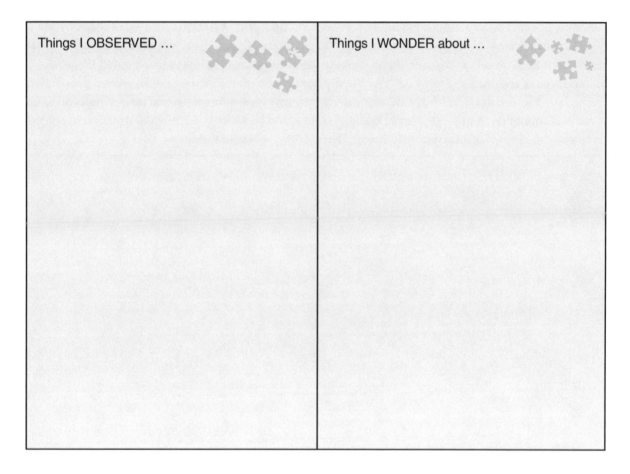

Things I OBSERVED …	Things I WONDER about …

Fossils provide clues about the traits of organisms that lived on Earth a long time ago. There are two main types of fossils. The first type of fossil is called a *body fossil*. Body fossils are the preserved remains of a plant or animal. The second type of fossil is called a *trace fossil*. Trace fossils are the remains of the activity of an animal. Trace fossils include footprints, imprints of shells or body parts, and nests. Fossils

can be found high on mountains, underwater, in the desert, on beaches, or underground. Scientists can use fossils to learn about different organisms that are no longer found on Earth and how the traits of different types of organisms have changed over time.

Fossils can also provide clues about what an ecosystem at a specific location was like in the past. An ecosystem includes all of the living things and non-living things in a given area. There are two main types of ecosystems. The first main type of ecosystem is called an *aquatic ecosystem.* An aquatic ecosystem is found in a body of water. Aquatic ecosystems are classified as either marine (ocean or sea) or freshwater (lake, marsh, river, or swamp). The second main type of ecosystem is called a *terrestrial ecosystem.* A terrestrial ecosystem is found on land. Terrestrial ecosystems are classified as forest, desert, grassland, or mountain.

Scientists can use fossils to learn about what an ecosystem was like in the past because organisms have specific traits that allow them to survive in a specific ecosystem. For example, fish, clams, and seaweed can survive in an aquatic ecosystem but not in a terrestrial one. Therefore, if a scientist finds a fossil fish in a desert, he or she can assume that there was once an ocean or lake at that location.

In this investigation you will examine several different fossils that were found near Darmstadt, Germany. These fossils are about 49 million years old. Your goal is to figure out what the ecosystem near Darmstadt was like 49 million years ago. To accomplish this task, you will need to make observations about these fossils to make inferences about the traits of the organisms that created those fossils and the type of environment in which they lived. Scientists can determine where an animal lived based on its traits because the structure of an organism determines how it functions and places limits on what it can and cannot do. You can therefore use the relationship between structure and function in animal bodies to determine what the ecosystem near Darmstadt was like in the past.

Things we KNOW from what we read …	What we will NEED to figure out …

Your Task

Use what you know about plants, animals, and the relationship between structure and function in nature to plan and carry out an investigation to determine what types of organisms lived near Darmstadt 49 million years ago and what the environment was like at that time.

The *guiding question* of this investigation is, **What was the ecosystem at Darmstadt like 49 million years ago?**

Materials

You may use any of the following materials during your investigation:

- Fossil A
- Fossil B
- Fossil C
- Fossil D
- Fossil E
- Fossil F

Safety Rules

Follow all normal safety rules.

Plan Your Investigation

Prepare a plan for your investigation by filling out the chart that follows; this plan is called an *investigation proposal*. Before you start developing your plan, be sure to discuss the following questions with the other members of your group:

- How might the **structure** of what you are studying relate to its **function?**
- What types of **patterns** might we look for to help answer the guiding question?

Investigation Log

Our guiding question:

We will collect the following data:

These are the steps we will follow to collect data:

I approve of this investigation proposal.

_____   _____
Teacher's signature   Date

National Science Teachers Association

Collect Your Data

Keep a record of what you measure or observe during your investigation in the space below.

Analyze Your Data

You will need to analyze the data you collected before you can develop an answer to the guiding question. In the space below, you can create a table or a graph or use pictures to show the traits or structures of the organisms found in the fossils.

Investigation Log

Draft Argument

Develop an argument on a whiteboard. It should include the following parts:

1. A *claim:* Your answer to the guiding question.
2. *Evidence:* An analysis of the data and an explanation of what the analysis means.
3. A *justification of the evidence:* Why your group thinks the evidence is important.

The Guiding Question:	
Our Claim:	
Our Evidence:	Our Justification of the Evidence:

Argumentation Session

Share your argument with your classmates. Be sure to ask them how to make your draft argument better. Keep track of their suggestions in the space below.

Ways to IMPROVE our argument …

Draft Report

Prepare an *investigation report* to share what you have learned. Use the information in this handout and your group's final argument to write a *draft* of your investigation report.

166

National Science Teachers Association

Introduction

We have been studying _____ in class. Before we started

this investigation, we explored _____

_____

_____

We noticed _____

_____

_____

My goal for this investigation was to figure out _____

_____

_____

The guiding question was _____

_____

_____

Method

To gather the data I needed to answer this question, I _____

_____

_____

_____

I then analyzed the data I collected by _____

_____

_____

_____

Investigation Log

Argument

My claim is _____

_____

_____

_____

_____

The _____ below shows _____

_____

_____

_____

This evidence is important because _____

_____

_____

_____

_____

_____

_____

_____

_____

_____

_____

 ### Review

Your friends need your help! Review the draft of their investigation reports and give them ideas about how to improve. Use the *peer-review guide* that begins on the next page to guide your review.

Investigation Log

Peer-Review Guide

Section 1: The Investigation	Reviewer Rating		
1. Did the author do a good job of explaining what the investigation was about?	☐ No	☐ Almost	☐ Yes
2. Did the author do a good job of making the **guiding question** clear?	☐ No	☐ Almost	☐ Yes
3. Did the author do a good job of describing what he or she did to **collect data?**	☐ No	☐ Almost	☐ Yes
4. Did the author do a good job describing **how** he or she **analyzed** the data?	☐ No	☐ Almost	☐ Yes
Reviewers: If your group gave the author any "No" or "Almost" ratings, please give the author some advice about what to do to improve this part of his or her investigation report.			

Section 2: The Argument	Reviewer Rating		
1. Does the author's claim provide a clear and detailed **answer** to the guiding question?	☐ No	☐ Almost	☐ Yes
2. Did the author support his or her claim with **scientific evidence?** Scientific evidence includes analyzed data and an explanation of the analysis.	☐ No	☐ Almost	☐ Yes
3. Does the **evidence** that the author uses in his or her argument **support the claim?**	☐ No	☐ Almost	☐ Yes
4. Did the author include enough **evidence** in his or her argument?	☐ No	☐ Almost	☐ Yes
5. Did the author do a good job of **explaining why the evidence** is important (why it matters)?	☐ No	☐ Almost	☐ Yes
6. Is the content of the argument **correct** based on the science concepts we talked about in class?	☐ No	☐ Almost	☐ Yes
Reviewers: If your group gave the author any "No" or "Almost" ratings, please give the author some advice about what to do to improve this part of his or her investigation report.			

Continued

National Science Teachers Association

Section 3: Mechanics	Reviewer Rating		
1. *Grammar:* Are the sentences complete? Is there proper subject-verb agreement in each sentence? Are there no run-on sentences?	☐ No	☐ Almost	☐ Yes
2. *Conventions:* Did the author use proper spelling, punctuation, and capitalization?	☐ No	☐ Almost	☐ Yes
3. *Word Choice:* Did the author use the right words in each sentence (for example, *there* vs. *their, to* vs. *too, then* vs. *than*)?	☐ No	☐ Almost	☐ Yes

Reviewers: If your group gave the author any "No" or "Almost" ratings, please give the author some advice about what to do to improve the writing mechanics of his or her investigation report.

General Reviewer Comments

We liked …

We wonder …

Write Your Final Report

Once you have received feedback from your friends about your draft report, create your final investigation report in the space that follows.

Introduction

_____

_____

_____

_____

_____

_____

_____

Method

_____

_____

_____

_____

_____

_____

_____

_____

_____

_____

_____

Argument

_____

_____

_____

_____

_____

_____

_____

_____

_____

_____

_____

_____

_____

Investigation Report Grading Rubric

Section 1: The Investigation	Score		
	Missing	Somewhat	Yes
1. The author explained what the investigation was about.	0	1	2
2. The author made the **guiding question** clear.	0	1	2
3. The author **described** what he or she did to **collect data.**	0	1	2
4. The author described **how** he or she **analyzed** the data.	0	1	2

Section 2: The Argument	Score		
	Missing	Somewhat	Yes
1. The claim includes a clear and detailed **answer** to the guiding question.	0	1	2
2. The author used **scientific evidence** to support the claim. Scientific evidence includes analyzed data and an explanation of the analysis.	0	1	2
3. The evidence **supports the claim.**	0	1	2
4. The author included enough **evidence** in his or her argument.	0	1	2
5. The author **explained why the evidence** is important.	0	1	2
6. The content of the argument is **correct.**	0	1	2

Section 3: Mechanics	Score		
	Missing	Somewhat	Yes
1. *Grammar:* The sentences are complete. There is proper subject-verb agreement in each sentence. There are no run-on sentences.	0	1	2
2. *Conventions:* The author used proper spelling, punctuation, and capitalization.	0	1	2
3. *Word Choice:* The author used the right words in each sentence (e.g., *there* vs. *their, to* vs. *too, then* vs. *than*).	0	1	2

Teacher Comments

Here are some things I really liked about your report …	Here are some things I think you could do next time to make your report even better …

Total: _____ /26

Checkout Questions

Investigation 10. Fossils: What Was the Ecosystem at Darmstadt Like 49 Million Years Ago?

1. Pictured below is the skeleton of an animal. What type of ecosystem do you think this animal lived in while it was alive? You may choose more than one ecosystem.

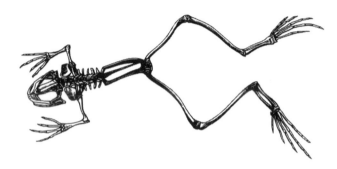

☐ Freshwater aquatic (lake, marsh, river, or swamp)

☐ Marine aquatic (ocean or sea)

☐ Terrestrial (forest, desert, grassland, or mountain)

2. Pictured below is the skeleton of an animal. What type of ecosystem do you think this animal lived in while it was alive? You may choose more than one ecosystem.

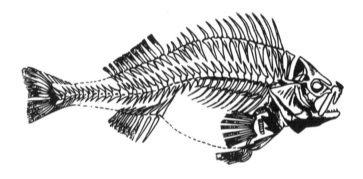

☐ Freshwater aquatic (lake, marsh, river, or swamp)

☐ Marine aquatic (ocean or sea)

☐ Terrestrial (forest, desert, grassland, or mountain)

Checkout Questions

3. Pictured below is the skeleton of an animal. What type of ecosystem do you think this animal lived in while it was alive? You may choose more than one ecosystem.

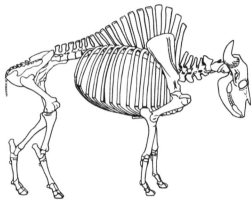

☐ Freshwater aquatic (lake, marsh, river, or swamp)

☐ Marine aquatic (ocean or sea)

☐ Terrestrial (forest, desert, grassland, or mountain)

4. Explain your thinking. How did the *structure* of the skeletons of these animals allow you to determine where they might have lived when they were alive?

_____

_____

_____

_____

_____

_____

Teacher Scoring Rubric for the Checkout Questions

Level	Description
3	The student can apply the core idea correctly in all cases and can fully explain the relationship between structure and function.
2	The student can apply the core idea correctly in all cases but cannot fully explain the relationship between structure and function.
1	The student cannot apply the core idea correctly in all cases but can fully explain the relationship between structure and function.
0	The student cannot apply the core idea correctly in all cases and cannot explain the relationship between structure and function.

National Science Teachers Association

Investigation 11

Differences in Traits: How Does Fur Color Affect the Likelihood That a Rabbit Will Survive?

Introduction

Animals and plants are adapted to the ecosystems where they live. They eat specific food and are often hunted by a specific predator that lives in the same ecosystem. They also have traits to help them survive. For example, animals who need to hunt at night will often have large ears and use sound to hunt their prey since it is hard to see in the dark. Other traits of animals may also make them more likely to survive.

Take some time to play with a simulation called *Natural Selection*. Keep track of what you observe and what you are wondering about in the boxes below.

Things I OBSERVED …	Things I WONDER about …

Investigation Log

Plants and animals are adapted to survive in their environments. Certain traits may help animals find and eat food, hide from predators, or survive harsh weather. For many animals, predators are a major threat to survival. Because of this, prey animals often have *advantageous* (helpful) traits that make them more difficult to see or to catch. These traits can keep prey animals from being discovered and eaten by predators. Plants can also protect themselves against predators. For example, many plants have sharp thorns that can poke animals that try to eat them. When a plant or animal is able to survive, it can reproduce to make new animals or plants. These new plants or animals will often have the same advantageous traits that their parents had. Over time, a whole species can develop these advantageous traits because the animals with those traits pass them on to their offspring, while animals with harmful or less advantageous traits die and are not able to reproduce.

In this investigation you need to figure out how a rabbit's fur color affects how likely it is to survive. To accomplish this task, you will need to use what you know about advantageous (helpful) and less advantageous (harmful) traits and an online simulation to collect data to determine how well rabbits with different colors of fur (a cause) are able to survive (the effect) in different types of environments. Scientists often look for cause-and-effect relationships like this to better understand how animals are able to survive in a specific ecosystem.

Things we KNOW from what we read …	What we will NEED to figure out …

Your Task

Use what you know about traits, ecosystems, and cause-and-effect relationships to design and carry out an investigation to determine how a rabbit's fur color affects the likelihood that the rabbit will survive.

The *guiding question* of this investigation is, ***How does fur color affect the likelihood that a rabbit will survive?***

Materials

You will use a computer or tablet and an online simulation called *Natural Selection* to conduct your investigation; the simulation is available at *https://phet.colorado.edu/en/ simulation/legacy/natural-selection.*

Safety Rules

Follow all normal safety rules.

Plan Your Investigation

Prepare a plan for your investigation by filling out the chart that follows; this plan is called an *investigation proposal.* Before you start developing your plan, be sure to discuss the following questions with the other members of your group:

- What types of **patterns** might we look for to help answer the guiding question?
- What information do we need to find a **cause-and-effect relationship?**

Our guiding question:

We will collect the following data:

These are the steps we will follow to collect data:

I approve of this investigation proposal.

_____   _____
Teacher's signature   Date

Collect Your Data

Keep a record of what you measure or observe during your investigation in the space below.

Analyze Your Data

You will need to analyze the data you collected before you can develop an answer to the guiding question. In the space below, create two graphs. One graph should show how many of each type of rabbit survived in the equatorial environment. The second graph should show how many of each type of rabbit survived in the arctic environment.

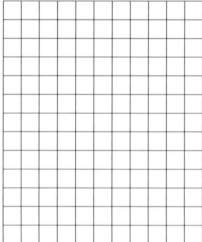

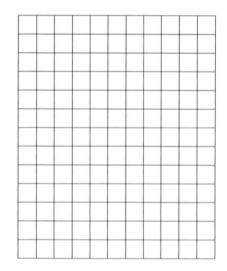

Draft Argument

Develop an argument on a whiteboard. It should include the following parts:

1. A *claim:* Your answer to the guiding question.

2. *Evidence:* An analysis of the data and an explanation of what the analysis means.

3. A *justification of the evidence:* Why your group thinks the evidence is important.

The Guiding Question:	
Our Claim:	
Our Evidence:	Our Justification of the Evidence:

Argumentation Session

Share your argument with your classmates. Be sure to ask them how to make your draft argument better. Keep track of their suggestions in the space below.

Ways to IMPROVE our argument …

Draft Report

Prepare an *investigation report* to share what you have learned. Use the information in this handout and your group's final argument to write a *draft* of your investigation report.

Introduction

We have been studying _____ in class. Before we started

this investigation, we explored _____

_____

_____

We noticed _____

_____

_____

My goal for this investigation was to figure out _____

_____

_____

The guiding question was _____

_____

_____

Method

To gather the data I needed to answer this question, I _____

_____

_____

_____

_____

_____

Investigation Log

I then analyzed the data I collected by _____

_____

_____

_____

Argument

My claim is _____

_____

_____

_____

_____

The graphs below show _____

_____

_____

_____

_____

_____

National Science Teachers Association

This evidence is important because _____

_____

_____

_____

_____

_____

_____

_____

_____

_____

_____

_____

 **Review**

Your friends need your help! Review the draft of their investigation reports and
give them ideas about how to improve. Use the *peer-review guide* that begins on the
next page to guide your review.

Investigation Log

Peer-Review Guide

Section 1: The Investigation	Reviewer Rating		
1. Did the author do a good job of explaining what the investigation was about?	☐ No	☐ Almost	☐ Yes
2. Did the author do a good job of making the **guiding question** clear?	☐ No	☐ Almost	☐ Yes
3. Did the author do a good job of describing what he or she did to **collect data?**	☐ No	☐ Almost	☐ Yes
4. Did the author do a good job describing **how** he or she **analyzed** the data?	☐ No	☐ Almost	☐ Yes
Reviewers: If your group gave the author any "No" or "Almost" ratings, please give the author some advice about what to do to improve this part of his or her investigation report.			

Section 2: The Argument	Reviewer Rating		
1. Does the author's claim provide a clear and detailed **answer** to the guiding question?	☐ No	☐ Almost	☐ Yes
2. Did the author support his or her claim with **scientific evidence?** Scientific evidence includes analyzed data and an explanation of the analysis.	☐ No	☐ Almost	☐ Yes
3. Does the **evidence** that the author uses in his or her argument **support the claim?**	☐ No	☐ Almost	☐ Yes
4. Did the author include enough **evidence** in his or her argument?	☐ No	☐ Almost	☐ Yes
5. Did the author do a good job of **explaining why the evidence** is important (why it matters)?	☐ No	☐ Almost	☐ Yes
6. Is the content of the argument **correct** based on the science concepts we talked about in class?	☐ No	☐ Almost	☐ Yes
Reviewers: If your group gave the author any "No" or "Almost" ratings, please give the author some advice about what to do to improve this part of his or her investigation report.			

Continued

National Science Teachers Association

Section 3: Mechanics	Reviewer Rating		
1. *Grammar:* Are the sentences complete? Is there proper subject-verb agreement in each sentence? Are there no run-on sentences?	☐ No	☐ Almost	☐ Yes
2. *Conventions:* Did the author use proper spelling, punctuation, and capitalization?	☐ No	☐ Almost	☐ Yes
3. *Word Choice:* Did the author use the right words in each sentence (for example, *there* vs. *their, to* vs. *too, then* vs. *than*)?	☐ No	☐ Almost	☐ Yes

Reviewers: If your group gave the author any "No" or "Almost" ratings, please give the author some advice about what to do to improve the writing mechanics of his or her investigation report.

General Reviewer Comments

We liked …

We wonder …

Write Your Final Report

Once you have received feedback from your friends about your draft report, create your final investigation report in the space that follows.

Introduction

_____

_____

_____

_____

_____

_____

_____

_____

Method

_____

_____

_____

_____

_____

_____

_____

_____

_____

_____

_____

Argument

_____

_____

_____

_____

_____

_____

_____

_____

_____

_____

_____

_____

_____

Investigation Log

Investigation Report Grading Rubric

Section 1: The Investigation	Score		
	Missing	Somewhat	Yes
1. The author explained what the investigation was about.	0	1	2
2. The author made the **guiding question** clear.	0	1	2
3. The author **described** what he or she did to **collect data.**	0	1	2
4. The author described **how** he or she **analyzed** the data.	0	1	2

Section 2: The Argument	Score		
	Missing	Somewhat	Yes
1. The claim includes a clear and detailed **answer** to the guiding question.	0	1	2
2. The author used **scientific evidence** to support the claim. Scientific evidence includes analyzed data and an explanation of the analysis.	0	1	2
3. The evidence **supports the claim.**	0	1	2
4. The author included enough **evidence** in his or her argument.	0	1	2
5. The author **explained why the evidence** is important.	0	1	2
6. The content of the argument is **correct.**	0	1	2

Section 3: Mechanics	Score		
	Missing	Somewhat	Yes
1. *Grammar:* The sentences are complete. There is proper subject-verb agreement in each sentence. There are no run-on sentences.	0	1	2
2. *Conventions:* The author used proper spelling, punctuation, and capitalization.	0	1	2
3. *Word Choice:* The author used the right words in each sentence (e.g., *there* vs. *their, to* vs. *too, then* vs. *than*).	0	1	2

Teacher Comments

Here are some things I really liked about your report …	Here are some things I think you could do next time to make your report even better …

Total: _____ /26

National Science Teachers Association

Investigation 11. Differences in Traits: How Does Fur Color Affect the Likelihood That a Rabbit Will Survive?

The picture below shows a population of mice that live in a sandy environment. Many hawks also live in this environment. Hawks eat mice.

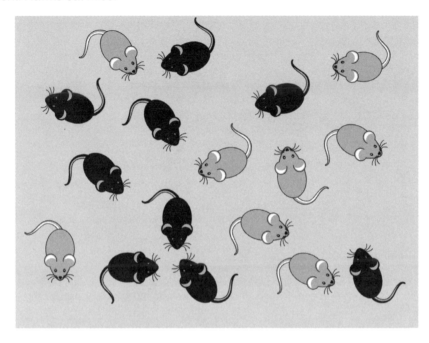

1. Which mice are most likely to become food for the hawks?

   A. Black mice

   B. Brown mice

   C. Both black and brown mice

2. Which fur color is most likely to become more common for mice in this environment over time (after many generations)?

   A. Black mice

   B. Brown mice

   C. Both black and brown mice

 **Checkout Questions**

Explain your thinking. What *cause-and-effect relationship* did you use to determine what would happen to this population of mice over time?

_____

_____

_____

_____

_____

_____

_____

_____

Teacher Scoring Rubric for the Checkout Questions

Level	Description
3	The student can apply the core idea correctly in all cases and can fully explain the cause-and-effect relationship.
2	The student can apply the core idea correctly in all cases but cannot fully explain the cause-and-effect relationship.
1	The student cannot apply the core idea correctly in all cases but can fully explain the cause-and-effect relationship.
0	The student cannot apply the core idea correctly in all cases and cannot explain the cause-and-effect relationship.

Investigation 12

Adaptations: Why Do Mammals That Live in the Arctic Ocean Have a Thick Layer of Blubber Under Their Skin?

Introduction

There are many different types of ecosystems on Earth. An ecosystem includes all of the living things and non-living things in a given area. The Arctic Ocean is an example of an aquatic ecosystem. Many different types of organisms live in the Arctic Ocean. Watch the video about the Arctic Ocean. As you watch, keep a record what you see (observe) and what you are wondering about in the boxes below.

Things I OBSERVED …	Things I WONDER about …

The Arctic Ocean is a cold and harsh environment. It is covered in ice most of the year. Some kinds of animals survive well in the Arctic Ocean, some survive less well, and some cannot survive at all. All the different types of animals that survive well there, such as bears, whales, birds, and fish, have specific traits that enable them to live in this specific environment. For example, some animals have white fur, feathers, or skin. These animals blend in with the ice that covers the Arctic Ocean. These white animals are able to either hide from predators or sneak up on their prey. Other traits help these animals deal with all the ice that covers the ocean. For example, the narwhal in the video you just watched has a long tusk that it can use to break through the ice that covers the ocean. Beluga whales are able to use sound to find their way around in the dark because it is often very dark under the ice that covers the ocean. All of these different traits make it possible for these animals to survive in the cold and harsh Arctic Ocean.

Sometimes, however, it is difficult to figure out why some animals have a specific trait. For example, all the mammals that live in the Arctic Ocean, such as bears, seals, and whales, have a thick layer of fat under their skin. This thick layer of fat is called *blubber*. In nature, the way an animal's body is shaped or structured determines how it functions and places limits on what it can or cannot do. The thick layer of blubber under the skin of mammals should therefore serve a function that helps these mammals survive in the Arctic Ocean in some way. One possible function is that a thick layer of blubber helps mammals stay warm when they are in very cold water. It is important for mammals that live in the Arctic Ocean to stay warm at all times because mammals will die if their body temperature drops too low.

In this investigation you will need to plan and carry out an experiment to test the hypothesis that a layer of blubber causes mammals to stay warm when they are in cold water. Scientists often investigate possible cause-and-effect relationships like this to explain how animals are able to survive in a specific environment.

Things we KNOW from what we read …	What we will NEED to figure out …

Your Task

Use what you know about the traits of animals, the characteristics of specific environments, cause-and-effect relationships, and how structure determines function in nature to test the idea that a thick layer of fat can help a mammal stay warm in a cold environment.

The *guiding question* of this investigation is, **Why do mammals that live in the Arctic Ocean have a thick layer of blubber under their skin?**

Materials

You may use any of the following materials during your investigation:

- Safety goggles (required)
- Plastic-bag glove filled with shortening (which is like blubber)
- Plastic-bag glove filled with feathers
- Plastic-bag glove (empty)
- Bucket of ice water
- 3 thermometers
- Stopwatch

Safety Rules

Follow all normal safety rules. In addition, be sure to follow these rules:

- Wear sanitized indirectly vented chemical-splash goggles during setup, investigation activity, and cleanup.
- Do not throw objects or put any objects in your mouth.
- Immediately wipe up any slip or fall hazards (such as water).
- Appropriately dispose of materials as directed by your teacher
- Wash your hands with soap and water when you are done collecting the data.

Plan Your Investigation

Prepare a plan for your investigation by filling out the chart that follows; this plan is called an *investigation proposal*. Before you start developing your plan, be sure to discuss the following questions with the other members of your group:

- What information do we need to find a **cause-and-effect relationship**?
- How might the **structure** of what you are studying relate to its **function**?

Our guiding question:

This is a picture of how we will set up the equipment:

We will collect the following data:

These are the steps we will follow to collect data:

I approve of this investigation proposal.

_____
Teacher's signature

_____
Date

Collect Your Data

Keep a record of what you measure or observe during your investigation in the space below.

Analyze Your Data

You will need to analyze the data you collected before you can develop an answer to the guiding question. To do this, create a graph that shows the relationship between the cause (what is inside the glove) and the effect (temperature inside the glove).

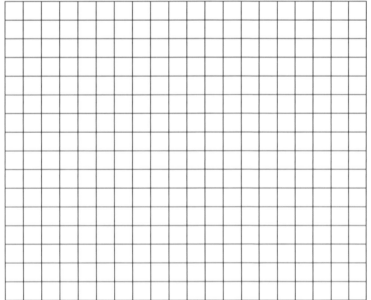

Draft Argument

Develop an argument on a whiteboard. It should include the following parts:

1. A *claim:* Your answer to the guiding question.
2. *Evidence:* An analysis of the data and an explanation of what the analysis means.
3. A *justification of the evidence:* Why your group thinks the evidence is important.

The Guiding Question:	
Our Claim:	
Our Evidence:	Our Justification of the Evidence:

Argumentation Session

Share your argument with your classmates. Be sure to ask them how to make your draft argument better. Keep track of their suggestions in the space below.

Ways to IMPROVE our argument ...

National Science Teachers Association

Draft Report

Prepare an *investigation report* to share what you have learned. Use the information in this handout and your group's final argument to write a *draft* of your investigation report.

Introduction

We have been studying _____ in class. Before we started

this investigation, we explored _____

_____

_____

We noticed _____

_____

_____

My goal for this investigation was to figure out _____

_____

_____

The guiding question was _____

_____

_____

Method

To gather the data I needed to answer this question, I _____

_____

_____

_____

_____

_____

Argument

My claim is _____

_____

_____

_____

_____

The graph below shows _____

_____

_____

_____

_____

_____

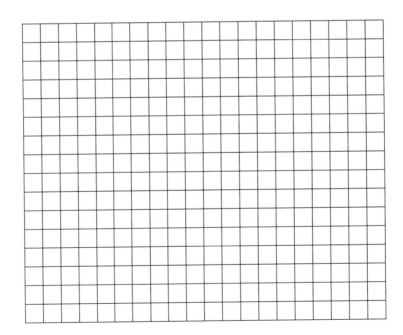

National Science Teachers Association

This evidence is important because _____

_____

_____

_____

_____

_____

_____

_____

_____

_____

 ### Review

Your friends need your help! Review the draft of their investigation reports and give them ideas about how to improve. Use the *peer-review guide* that begins on the next page to guide your review.

Investigation Log

Peer-Review Guide

Section 1: The Investigation	Reviewer Rating		
1. Did the author do a good job of explaining what the investigation was about?	☐ No	☐ Almost	☐ Yes
2. Did the author do a good job of making the **guiding question** clear?	☐ No	☐ Almost	☐ Yes
3. Did the author do a good job of describing what he or she did to **collect data?**	☐ No	☐ Almost	☐ Yes
4. Did the author do a good job describing **how** he or she **analyzed** the data?	☐ No	☐ Almost	☐ Yes
Reviewers: If your group gave the author any "No" or "Almost" ratings, please give the author some advice about what to do to improve this part of his or her investigation report.			

Section 2: The Argument	Reviewer Rating		
1. Does the author's claim provide a clear and detailed **answer** to the guiding question?	☐ No	☐ Almost	☐ Yes
2. Did the author support his or her claim with **scientific evidence?** Scientific evidence includes analyzed data and an explanation of the analysis.	☐ No	☐ Almost	☐ Yes
3. Does the **evidence** that the author uses in his or her argument **support the claim?**	☐ No	☐ Almost	☐ Yes
4. Did the author include enough **evidence** in his or her argument?	☐ No	☐ Almost	☐ Yes
5. Did the author do a good job of **explaining why the evidence** is important (why it matters)?	☐ No	☐ Almost	☐ Yes
6. Is the content of the argument **correct** based on the science concepts we talked about in class?	☐ No	☐ Almost	☐ Yes
Reviewers: If your group gave the author any "No" or "Almost" ratings, please give the author some advice about what to do to improve this part of his or her investigation report.			

Continued

National Science Teachers Association

Section 3: Mechanics	Reviewer Rating		
1. *Grammar:* Are the sentences complete? Is there proper subject-verb agreement in each sentence? Are there no run-on sentences?	☐ No	☐ Almost	☐ Yes
2. *Conventions:* Did the author use proper spelling, punctuation, and capitalization?	☐ No	☐ Almost	☐ Yes
3. *Word Choice:* Did the author use the right words in each sentence (for example, *there* vs. *their, to* vs. *too, then* vs. *than*)?	☐ No	☐ Almost	☐ Yes

Reviewers: If your group gave the author any "No" or "Almost" ratings, please give the author some advice about what to do to improve the writing mechanics of his or her investigation report.

General Reviewer Comments

We liked …

We wonder …

Write Your Final Report

Once you have received feedback from your friends about your draft report, create your final investigation report in the space that follows.

Introduction

_____

_____

_____

_____

_____

_____

_____

Method

_____

_____

_____

_____

_____

_____

_____

_____

_____

_____

Argument

_____

_____

_____

_____

_____

_____

_____

_____

_____

_____

_____

_____

Investigation Log

Investigation Report Grading Rubric

Section 1: The Investigation	Score		
	Missing	Somewhat	Yes
1. The author explained what the investigation was about.	0	1	2
2. The author made the **guiding question** clear.	0	1	2
3. The author **described** what he or she did to **collect data.**	0	1	2
4. The author described **how** he or she **analyzed** the data.	0	1	2

Section 2: The Argument	Score		
	Missing	Somewhat	Yes
1. The claim includes a clear and detailed **answer** to the guiding question.	0	1	2
2. The author used **scientific evidence** to support the claim. Scientific evidence includes analyzed data and an explanation of the analysis.	0	1	2
3. The evidence **supports the claim.**	0	1	2
4. The author included enough **evidence** in his or her argument.	0	1	2
5. The author **explained why the evidence** is important.	0	1	2
6. The content of the argument is **correct.**	0	1	2

Section 3: Mechanics	Score		
	Missing	Somewhat	Yes
1. *Grammar:* The sentences are complete. There is proper subject-verb agreement in each sentence. There are no run-on sentences.	0	1	2
2. *Conventions:* The author used proper spelling, punctuation, and capitalization.	0	1	2
3. *Word Choice:* The author used the right words in each sentence (e.g., *there* vs. *their, to* vs. *too, then* vs. *than*).	0	1	2

Teacher Comments	
Here are some things I really liked about your report …	Here are some things I think you could do next time to make your report even better …

Total: _____ /26

Checkout Questions

Investigation 12. Adaptations: Why Do Mammals That Live in the Arctic Ocean Have a Thick Layer of Blubber Under Their Skin?

Pictured below are two different types of animals. One is called a musk ox and the other is called a gazelle. The animals live in different parts of the world.

Musk ox

Gazelle

1. Which animal do you think is more likely to live in a cold environment?

   ☐ Musk ox   ☐ Gazelle

2. Explain your thinking. How did your understanding of the relationship between *cause and effect* or *structure and function* allow you to determine where the animal is most likely to live?

_____

_____

_____

_____

_____

_____

Teacher Scoring Rubric for the Checkout Questions

Level	Description
3	The student can apply the core idea correctly in all cases and can fully explain the relationship between cause and effect or structure and function.
2	The student can apply the core idea correctly in all cases but cannot fully explain the relationship between cause and effect or structure and function.
1	The student cannot apply the core idea correctly in all cases but can fully explain the relationship between cause and effect or structure and function.
0	The student cannot apply the core idea correctly in all cases and cannot explain the relationship between cause and effect or structure and function.

Section 6
Earth's Systems

Investigation 13

Weather Patterns: What Weather Conditions Can We Expect Here During Each Season?

Introduction

Weather is the current condition of the atmosphere at a specific location. You can also think of weather as what we see outside on a particular day. For example, it may be 75 degrees and sunny outside, or it may be 45 degrees and raining. Weather is important because many of us make plans about where we will go and what we will wear based on current weather conditions. Over the next week, you will have a chance to track how the weather conditions do or do not change. Keep track of what you observe each day and what you are wondering about in the boxes below.

Things I observed on …				
Day 1	Day 2	Day 3	Day 4	Day 5

Things I WONDER about …

Meteorologists are people who study and predict the weather. A meteorologist will often describe the current weather conditions using several different measurements: temperature, humidity, wind speed, precipitation, and cloud cover.

- *Temperature:* Temperature is measured in degrees Fahrenheit or degrees Celsius and tells us how hot or how cold it is outside.

- *Humidity:* Humidity tells us how much water is in the air and is often reported as a percentage (for example, 50%). When it is very humid outside, the air feels wet. People often describe a very humid day as being "muggy" or "sticky."

- *Wind speed:* Wind is the movement of air from one spot to another. Wind is measured in miles per hour (mph). During a major category 5 hurricane, wind speeds can reach over 160 mph!

- *Precipitation:* Precipitation is the amount of rain, snow, sleet, or hail that falls over a specific time period. It is usually reported in millimeters or inches.

- *Cloud cover:* The amount of clouds in the sky can be described as clear (no clouds in the sky), partly cloudy (less than half cloud cover), mostly cloudy (more than half cloud cover but with some breaks), or overcast (complete cloud cover).

Meteorologists and other scientists record patterns of the weather over time and at different locations so that they can make predictions about what kind of weather to expect during each season. This is important to know because people often want to know what the weather tends to be like in different cities during each season so they know what types of plants to grow, when to plant them, and what type of clothes to buy. If someone just moved to your city, they might ask you questions like, "Do you get a lot of rain here?" or "Is it really hot during the summer?" In this investigation you will have an opportunity to figure out what kind of weather conditions can be expected where you live during each season.

Things we KNOW from what we read …	What we will NEED to figure out …

National Science Teachers Association

Your Task

Use information about the daily weather conditions over the last year and what you know about the weather, seasons, and patterns to determine the typical weather that can be expected at your school during each season. Spring starts on March 21 and ends on June 20. Summer begins on June 21 and ends on September 20. Fall begins on September 21 and ends on December 20. Winter starts on December 21 and ends on March 20. Be sure to pick at least two measurements to describe the daily weather conditions at your school.

The *guiding question* of this investigation is, ***What weather conditions can we expect here during each season?***

Materials

You may use any of the following materials during your investigation:

- Computer or tablet with internet access
- Thermometer (optional)
- Barometer (optional)
- Hygrometer (optional)
- Anemometer (optional)
- Weather app for smartphone or tablet (optional)

You will need to access a website called World Weather and Climate Information. The website is at *https://weather-and-climate.com.*

Safety Rules

Follow all normal safety rules.

Plan Your Investigation

Prepare a plan for your investigation by filling out the chart that follows; this plan is called an *investigation proposal.* Before you start developing your plan, be sure to discuss the following questions with the other members of your group:

- What information should we collect so we can **describe** the typical weather here?
- What types of **patterns** might we look for to help answer the guiding question?

Our guiding question:

We will collect the following data:

These are the steps we will follow to collect data:

I approve of this investigation proposal.

_____   _____
Teacher's signature   Date

Collect Your Data

Keep a record of what you measure or observe during your investigation in the space below.

Analyze Your Data

You will need to analyze the data you decided to use before you can develop an answer to the guiding question. In the space that follows, create two graphs. Each graph should show how a different weather measurement changes by season where you live.

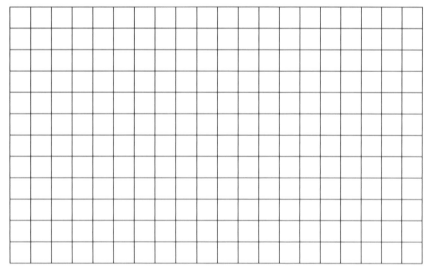

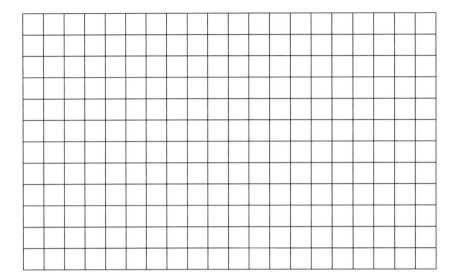

Draft Argument

Develop an argument on a whiteboard. It should include the following parts:

1. A *claim:* Your answer to the guiding question.

2. *Evidence:* An analysis of the data and an explanation of what the analysis means.

3. A *justification of the evidence:* Why your group thinks the evidence is important.

The Guiding Question:	
Our Claim:	
Our Evidence:	Our Justification of the Evidence:

Argumentation Session

Share your argument with your classmates. Be sure to ask them how to make your draft argument better. Keep track of their suggestions in the space below.

Ways to IMPROVE our argument …

Draft Report

Prepare an *investigation report* to share what you have learned. Use the information in this handout and your group's final argument to write a *draft* of your investigation report.

Introduction

We have been studying _____ in class. Before we

started this investigation, we explored _____

_____

_____

We noticed _____

_____

_____

My goal for this investigation was to figure out _____

_____

_____

The guiding question was _____

_____

_____

Method

To gather the data I needed to answer this question, I _____

_____

_____

_____

_____

_____

Investigation Log

I then analyzed the data I collected by _____

_____

_____

_____

_____

Argument

My claim is _____

_____

_____

_____

Figure 1 below shows _____

_____

_____

_____

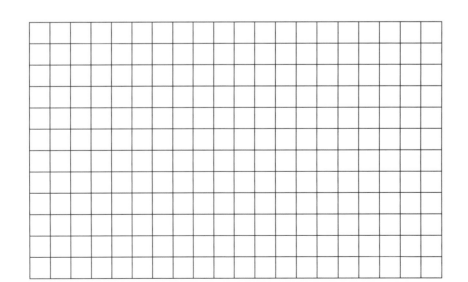

Figure 2 below shows _____

_____

_____

_____

This evidence is important because _____

_____

_____

_____

Review

Your friends need your help! Review the draft of their investigation reports and give them ideas about how to improve. Use the *peer-review guide* that begins on the next page to guide your review.

Investigation Log

Peer-Review Guide

Section 1: The Investigation	Reviewer Rating		
1. Did the author do a good job of explaining what the investigation was about?	☐ No	☐ Almost	☐ Yes
2. Did the author do a good job of making the **guiding question** clear?	☐ No	☐ Almost	☐ Yes
3. Did the author do a good job of describing what he or she did to **collect data?**	☐ No	☐ Almost	☐ Yes
4. Did the author do a good job describing **how** he or she **analyzed** the data?	☐ No	☐ Almost	☐ Yes

Reviewers: If your group gave the author any "No" or "Almost" ratings, please give the author some advice about what to do to improve this part of his or her investigation report.

Section 2: The Argument	Reviewer Rating		
1. Does the author's claim provide a clear and detailed **answer** to the guiding question?	☐ No	☐ Almost	☐ Yes
2. Did the author support his or her claim with **scientific evidence?** Scientific evidence includes analyzed data and an explanation of the analysis.	☐ No	☐ Almost	☐ Yes
3. Does the **evidence** that the author uses in his or her argument **support the claim?**	☐ No	☐ Almost	☐ Yes
4. Did the author include enough **evidence** in his or her argument?	☐ No	☐ Almost	☐ Yes
5. Did the author do a good job of **explaining why the evidence** is important (why it matters)?	☐ No	☐ Almost	☐ Yes
6. Is the content of the argument **correct** based on the science concepts we talked about in class?	☐ No	☐ Almost	☐ Yes

Reviewers: If your group gave the author any "No" or "Almost" ratings, please give the author some advice about what to do to improve this part of his or her investigation report.

Continued

National Science Teachers Association

Section 3: Mechanics	Reviewer Rating		
1. *Grammar:* Are the sentences complete? Is there proper subject-verb agreement in each sentence? Are there no run-on sentences?	☐ No	☐ Almost	☐ Yes
2. *Conventions:* Did the author use proper spelling, punctuation, and capitalization?	☐ No	☐ Almost	☐ Yes
3. *Word Choice:* Did the author use the right words in each sentence (for example, *there* vs. *their, to* vs. *too, then* vs. *than*)?	☐ No	☐ Almost	☐ Yes

Reviewers: If your group gave the author any "No" or "Almost" ratings, please give the author some advice about what to do to improve the writing mechanics of his or her investigation report.

General Reviewer Comments

We liked …

We wonder …

Write Your Final Report

Once you have received feedback from your friends about your draft report, create your final investigation report in the space that follows.

Introduction

_____

_____

_____

_____

_____

_____

_____

Method

_____

_____

_____

_____

_____

_____

_____

_____

_____

_____

_____

National Science Teachers Association

Argument

_____

_____

_____

_____

_____

_____

_____

_____

_____

_____

_____

_____

_____

Investigation Log

Investigation Report Grading Rubric

Section 1: The Investigation	Missing	Somewhat	Yes
		Score	
1. The author explained what the investigation was about.	0	1	2
2. The author made the **guiding question** clear.	0	1	2
3. The author **described** what he or she did to **collect data.**	0	1	2
4. The author described **how** he or she **analyzed** the data.	0	1	2

Section 2: The Argument	Missing	Somewhat	Yes
		Score	
1. The claim includes a clear and detailed **answer** to the guiding question.	0	1	2
2. The author used **scientific evidence** to support the claim. Scientific evidence includes analyzed data and an explanation of the analysis.	0	1	2
3. The evidence **supports the claim.**	0	1	2
4. The author included enough **evidence** in his or her argument.	0	1	2
5. The author **explained why the evidence** is important.	0	1	2
6. The content of the argument is **correct.**	0	1	2

Section 3: Mechanics	Missing	Somewhat	Yes
		Score	
1. *Grammar:* The sentences are complete. There is proper subject-verb agreement in each sentence. There are no run-on sentences.	0	1	2
2. *Conventions:* The author used proper spelling, punctuation, and capitalization.	0	1	2
3. *Word Choice:* The author used the right words in each sentence (e.g., *there* vs. *their, to* vs. *too, then* vs. *than*).	0	1	2

Teacher Comments

Here are some things I really liked about your report …	Here are some things I think you could do next time to make your report even better …

Total: _____ /26

Checkout Questions

Investigation 13. Weather Patterns: What Weather Conditions Can We Expect Here During Each Season?

The graph below shows the average amount of rainfall in Austin, Texas, by month. This graph is missing the amount of average rainfall for the months of August and October.

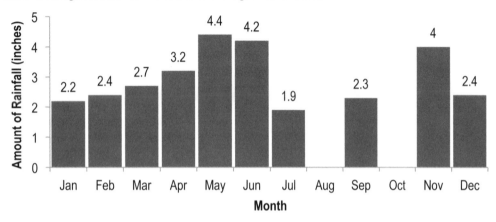

1. How much rainfall would you expect to see in Austin during the month of August?

   _____

2. How much rainfall would you expect to see in Austin during the month of October?

   _____

3. Explain your thinking. What *pattern* from your investigation did you use to predict the amount of rainfall in Austin during August and October?

   _____

   _____

   _____

   _____

   _____

Teacher Scoring Rubric for the Checkout Questions

Level	Description
3	The student can apply the core idea correctly in all cases and can fully explain the pattern.
2	The student can apply the core idea correctly in all cases but cannot fully explain the pattern.
1	The student cannot apply the core idea correctly in all cases but can fully explain the pattern.
0	The student cannot apply the core idea correctly in all cases and cannot explain the pattern.

Investigation 14

Climate and Location: How Does the Climate Change as One Moves From the Equator Toward the Poles?

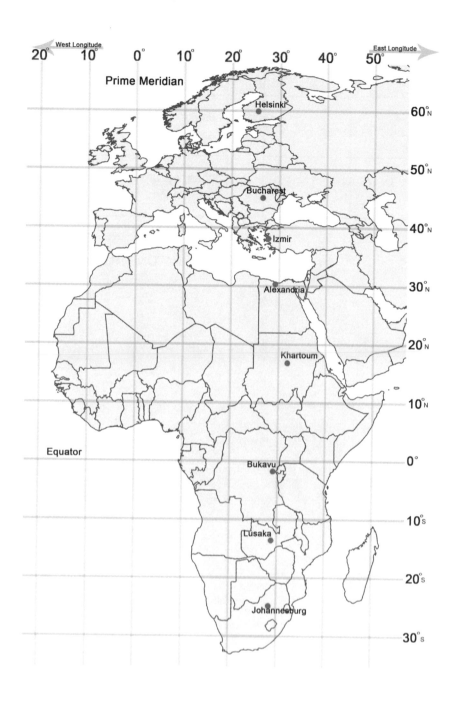

Investigation Log

Introduction

There are many cities all over the world. All of these cities have characteristics that make them special. Take a minute to find the latitude and longitude of the eight different cities labeled on the map on the previous page. Keep track of what you notice and what you are wondering about in the boxes below.

City	Longitude	Latitude
Helsinki, Finland		
Bucharest, Romania		
Izmir, Turkey		
Alexandria, Egypt		

City	Longitude	Latitude
Khartoum, Sudan		
Bukavu, Democratic Republic of the Congo		
Lusaka, Zambia		
Johannesburg, South Africa		

Things I NOTICED ...

Things I WONDER about ...

National Science Teachers Association

People often want to know about the climate and the current weather in a city before they travel to that city so they know what type of clothing to bring with them. *Climate* is a pattern of weather in a particular region over a long period of time. *Weather* is the current condition of the atmosphere at a specific place. We describe the weather by measuring the air temperature, humidity, wind speed, precipitation, and cloud cover. Weather can change from hour to hour or day to day.

A region's weather patterns, tracked for more than 30 years, are used to describe the climate of that region. There are five main climate types:

- *Tropical*—a region that is warm all year and gets a lot of rain
- *Dry*—a region that gets very little rain; it can be hot or cool
- *Mild*—a region with warm and dry summers and short, cool, but rainy winters
- *Continental*—a region with short summers and long winters that are cold with a lot of snow
- *Polar*—a region with temperatures that are cold all year

There are many reasons why different regions have different climates. One cause that may or may not affect the climate of a specific region is *latitude,* or how far that specific region is from the equator. Think about the eight cities shown on the map. These cities are all located at between about 25 and 30 East longitude, but each city is at a different latitude. *Longitude* is the distance east or west of the *prime meridian* (an imaginary line running from north to south through Greenwich, England). Some of the cities, in other words, are close to the equator and some are far away from the equator, even though all these cities are found on the same side of the Earth.

Your goal in this investigation is to first determine if these cities have different climates and then use this information to figure out if the climate at a specific location is related to how far it is from the equator. To accomplish this task, you will need to compare how the typical weather in at least two of these cities changes by month over an entire year. You can then use this information to look for a pattern. If you can find a pattern, you will be able to figure out how latitude and climate are related.

Things we KNOW from what we read …	What we will NEED to figure out …

Your Task

Use what you know about weather, climate, and patterns to determine the climate of at least two different cities that are located at a similar longitude but different latitudes. Then determine if there is a relationship between latitude and climate.

The *guiding question* of this investigation is, **How does the climate change as one moves from the equator toward the poles?**

Materials

You will use a computer or tablet with internet access and a website called World Weather and Climate Information during your investigation. The website is at *https://weather-and-climate.com*.

Safety Rules

Follow all normal safety rules.

Plan Your Investigation

Prepare a plan for your investigation by filling out the chart that follows; this plan is called an *investigation proposal*. Before you start developing your plan, be sure to discuss the following questions with the other members of your group:

- What information should we collect so we can **describe** the climate of a city?
- What types of **patterns** might we look for to help answer the guiding question?

Our guiding question:

We will collect the following data:

These are the steps we will follow to collect data:

I approve of this investigation proposal.

_____   _____
Teacher's signature   Date

Collect Your Data

Keep a record of what you measure or observe during your investigation in the space below.

Analyze Your Data

You will need to analyze the data you collected before you can develop an answer to the guiding question. In the space that follows, create two graphs to illustrate how the climate of each city is different.

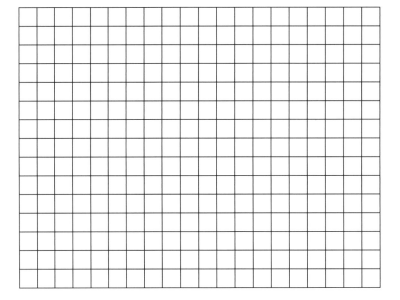

National Science Teachers Association

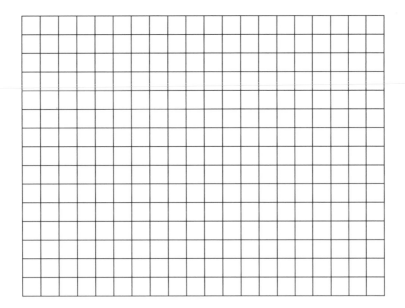

Draft Argument

Develop an argument on a whiteboard. It should include the following parts:

1. A *claim:* Your answer to the guiding question.

2. *Evidence:* An analysis of the data and an explanation of what the analysis means.

3. A *justification of the evidence:* Why your group thinks the evidence is important.

The Guiding Question:	
Our Claim:	
Our Evidence:	Our Justification of the Evidence:

Argumentation Session

Share your argument with your classmates. Be sure to ask them how to make your draft argument better. Keep track of their suggestions in the space below.

Ways to IMPROVE our argument …

Draft Report

Prepare an *investigation report* to share what you have learned. Use the information in this handout and your group's final argument to write a *draft* of your investigation report.

Introduction

We have been studying _____ in class. Before we

started this investigation, we explored _____

_____

_____

We noticed _____

_____

_____

My goal for this investigation was to figure out _____

_____

_____

The guiding question was _____

_____

_____

Method

To gather the data I needed to answer this question, I _____

_____

_____

_____

_____

_____

_____

_____

Investigation Log

I then analyzed the data I collected by _____

_____

_____

_____

Argument

My claim is _____

_____

_____

_____

Figure 1 below shows _____

_____

_____

_____

_____

_____

_____

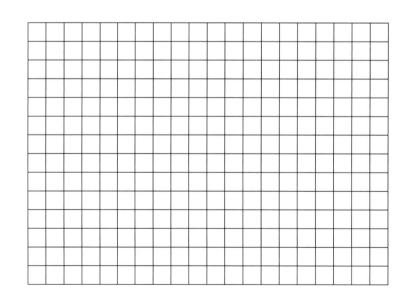

Figure 2 below shows _____

_____

_____

_____

_____

_____

_____

This evidence is important because _____

_____

_____

_____

_____

_____

_____

_____

_____

Review

Your friends need your help! Review the draft of their investigation reports and give them ideas about how to improve. Use the *peer-review guide* that begins on the next page to guide your review.

Peer-Review Guide

Section 1: The Investigation	Reviewer Rating		
1. Did the author do a good job of explaining what the investigation was about?	☐ No	☐ Almost	☐ Yes
2. Did the author do a good job of making the **guiding question** clear?	☐ No	☐ Almost	☐ Yes
3. Did the author do a good job of describing what he or she did to **collect data?**	☐ No	☐ Almost	☐ Yes
4. Did the author do a good job describing **how** he or she **analyzed** the data?	☐ No	☐ Almost	☐ Yes
Reviewers: If your group gave the author any "No" or "Almost" ratings, please give the author some advice about what to do to improve this part of his or her investigation report.			

Section 2: The Argument	Reviewer Rating		
1. Does the author's claim provide a clear and detailed **answer** to the guiding question?	☐ No	☐ Almost	☐ Yes
2. Did the author support his or her claim with **scientific evidence?** Scientific evidence includes analyzed data and an explanation of the analysis.	☐ No	☐ Almost	☐ Yes
3. Does the **evidence** that the author uses in his or her argument **support the claim?**	☐ No	☐ Almost	☐ Yes
4. Did the author include enough **evidence** in his or her argument?	☐ No	☐ Almost	☐ Yes
5. Did the author do a good job of **explaining why the evidence** is important (why it matters)?	☐ No	☐ Almost	☐ Yes
6. Is the content of the argument **correct** based on the science concepts we talked about in class?	☐ No	☐ Almost	☐ Yes
Reviewers: If your group gave the author any "No" or "Almost" ratings, please give the author some advice about what to do to improve this part of his or her investigation report.			

Continued

Section 3: Mechanics	Reviewer Rating		
1. *Grammar:* Are the sentences complete? Is there proper subject-verb agreement in each sentence? Are there no run-on sentences?	☐ No	☐ Almost	☐ Yes
2. *Conventions:* Did the author use proper spelling, punctuation, and capitalization?	☐ No	☐ Almost	☐ Yes
3. *Word Choice:* Did the author use the right words in each sentence (for example, *there* vs. *their, to* vs. *too, then* vs. *than*)?	☐ No	☐ Almost	☐ Yes

Reviewers: If your group gave the author any "No" or "Almost" ratings, please give the author some advice about what to do to improve the writing mechanics of his or her investigation report.

General Reviewer Comments

We liked …

We wonder …

Write Your Final Report

Once you have received feedback from your friends about your draft report, create your final investigation report in the space that follows.

Introduction

_____

_____

_____

_____

_____

_____

_____

Method

_____

_____

_____

_____

_____

_____

_____

_____

_____

_____

_____

Investigation Log

Argument

_____

_____

_____

_____

_____

_____

_____

_____

_____

_____

_____

_____

_____

National Science Teachers Association

Investigation Report Grading Rubric

Section 1: The Investigation	Score		
	Missing	Somewhat	Yes
1. The author explained what the investigation was about.	0	1	2
2. The author made the **guiding question** clear.	0	1	2
3. The author **described** what he or she did to **collect data.**	0	1	2
4. The author described **how** he or she **analyzed** the data.	0	1	2

Section 2: The Argument	Score		
	Missing	Somewhat	Yes
1. The claim includes a clear and detailed **answer** to the guiding question.	0	1	2
2. The author used **scientific evidence** to support the claim. Scientific evidence includes analyzed data and an explanation of the analysis.	0	1	2
3. The evidence **supports the claim.**	0	1	2
4. The author included enough **evidence** in his or her argument.	0	1	2
5. The author **explained why the evidence** is important.	0	1	2
6. The content of the argument is **correct.**	0	1	2

Section 3: Mechanics	Score		
	Missing	Somewhat	Yes
1. *Grammar:* The sentences are complete. There is proper subject-verb agreement in each sentence. There are no run-on sentences.	0	1	2
2. *Conventions:* The author used proper spelling, punctuation, and capitalization.	0	1	2
3. *Word Choice:* The author used the right words in each sentence (e.g., *there* vs. *their, to* vs. *too, then* vs. *than*).	0	1	2

Teacher Comments	
Here are some things I really liked about your report …	Here are some things I think you could do next time to make your report even better …

Total: _____ /26

Checkout Questions

Investigation 14. Climate and Location: How Does the Climate Change as One Moves From the Equator Toward the Poles?

The graph below shows the average high temperature by month in two different cities. Both cities are located on the same line of longitude in the Western Hemisphere.

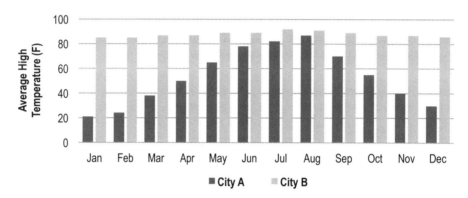

1. Which city has the greatest change in seasonal temperature?

   ☐ City A   ☐ City B

2. Which city is most likely located farthest from the equator?

   ☐ City A   ☐ City B

The graph below shows the average amount of precipitation by month in these two cities.

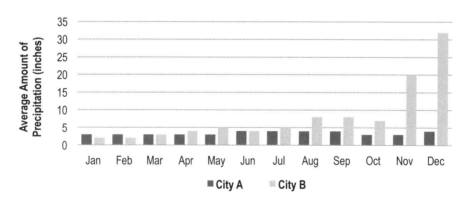

3. How would you describe the climate in city A based on all the information available?

   ☐ Tropical   ☐ Continental

   ☐ Dry   ☐ Polar

   ☐ Mild

National Science Teachers Association

4. How would you describe the climate in city B based on all the information available?

☐ Tropical   ☐ Continental

☐ Dry   ☐ Polar

☐ Mild

5. Explain your thinking. What *pattern* from your investigation did you use to determine the location and climate of these two cities?

_____

_____

_____

_____

_____

_____

_____

_____

_____

_____

Teacher Scoring Rubric for the Checkout Questions

Level	Description
3	The student can apply the core idea correctly in all cases and can fully explain the pattern.
2	The student can apply the core idea correctly in all cases but cannot fully explain the pattern.
1	The student cannot apply the core idea correctly in all cases but can fully explain the pattern.
0	The student cannot apply the core idea correctly in all cases and cannot explain the pattern.

IMAGE CREDITS

All images in this book are stock photographs or courtesy of the authors unless otherwise noted below.

Investigation 7

Earthworm illustration in checkout question 1: Modified from Pearson Scott Foresman, Wikimedia Commons, Public domain. *https://commons.wikimedia.org/wiki/File:Earthworm_1_(PSF).png*

Ladybug illustration in checkout question 2: Modified from Pearson Scott Foresman, Wikimedia Commons, Public domain. *https://commons.wikimedia.org/wiki/File:Ladybug_(PSF).svg*

Investigation 10

Frog skeleton illustration in checkout question 1: *https://pixabay.com/en/amphibian-animal-bone-dead-frog-2028309*

Investigation 12

Musk ox image above checkout question 1: Jeangagnon, Wikimedia Commons, CC BY-SA 2.0, *https://commons.wikimedia.org/wiki/File:Musk_ox.jpg*

Gazelle image above checkout question 1: Charles J. Sharp, Wikimedia Commons, CC BY-SA 3.0, *https://commons.wikimedia.org/wiki/File:Mountain_gazelle_(gazella_gazella).jpg*